T0136755

Intensive Agriculture and Sustainability

The Sustainability and the Environment series provides a comprehensive, independent, and critical evaluation of environmental and sustainability issues affecting Canada and the world today.

Anthony Scott, John Robinson, and David Cohen, eds., *Managing Natural Resources in British Columbia: Markets, Regulations, and Sustainable Development* (1995)

John B. Robinson, *Life in 2030: Exploring a Sustainable Future for Canada* (1996)

Ann Dale and John B. Robinson, eds., *Achieving Sustainable Development* (1996)

John T. Pierce and Ann Dale, eds., *Communities, Development, and Sustainability across Canada* (1999)

Robert F. Woollard and Aleck Ostry, eds., *Fatal Consumption: Rethinking Sustainable Development* (2000)

Ann Dale, *At the Edge: Sustainable Development in the 21st Century* (2001)

Mark Jaccard, John Nyboer, and Bryn Sadownik, *The Cost of Climate Policy* (2002)

Mike Carr, *Bioregionalism and Civil Society: Democratic Challenges to Corporate Globalism* (2004)

Glen Filson, ed., *Intensive Agriculture and Sustainability: A Farming Systems Analysis* (2004)

Sustainability
and the
Environment

Edited by Glen C. Filson

Intensive Agriculture and Sustainability: A Farming Systems Analysis

15 14 13 12 11 10 09 08 07 06 05 04   5 4 3 2 1

Printed in Canada on acid-free paper

Library and Archives Canada Cataloguing in Publication

Intensive agriculture and sustainability: a farming systems analysis / edited by Glen C. Filson.

(Sustainability and the environment)
Includes bibliographical references and index.
ISBN 0-7748-1104-8

1. Agricultural productivity – Environmental aspects – Ontario. 2. Livestock factories – Environmental aspects – Ontario. 3. Livestock factories – Social aspects – Ontario. 4. Agricultural systems – Research – Ontario. 5. Sustainable agriculture – Ontario. I. Filson, Glen C., 1947- II. Series.

S589.7.I58 2004   338.1'09713   C2004-904910-0

Canada

UBC Press gratefully acknowledges the financial support for our publishing program of the Government of Canada through the Book Publishing Industry Development Program (BPIDP), and of the Canada Council for the Arts, and the British Columbia Arts Council.

This book has been published with the help of a grant from the Canadian Federation for the Humanities and Social Sciences, through the Aid to Scholarly Publications Programme, using funds provided by the Social Sciences and Humanities Research Council of Canada.

UBC Press
The University of British Columbia
2029 West Mall
Vancouver, BC V6T 1Z2
604-822-5959 / Fax: 604-822-6083
www.ubcpress.ca

Contents

Figures and Tables

Figures

Tables

Foreword

Murray H. Miller

Farming systems research (FSR) gained prominence during the 1980s as an approach to improving the functioning of small-scale farms in developing countries. The focus has been primarily on increasing productivity and equitability through on-farm research and extension to improve farmers' management decisions. The FSR program at the University of Guelph differed in that it attempted to apply a holistic, integrated approach to understanding the functioning of intensive agricultural systems. The program evolved out of concern for the sustainability of intensive agriculture and for the impact of agricultural practices on the quality of soil, water, and air.

Agricultural research at the university had been managed under programs such as corn and oilseed production, soil management, beef production, pest management, agricultural economics, and so on. This led naturally to discipline-oriented research and a reductionist approach. Although this approach has brought tremendous improvements in agricultural production, it does not address the question of sustainability effectively. An ad hoc group consisting of crop, soil, and animal scientists, an environmental biologist, an agricultural economist, and a rural extension specialist was charged with the task of proposing a mechanism for achieving an integrated approach to research on sustainable agriculture.

Our first task was to define what, for our purposes, was a sustainable agricultural system. After considerable discussion, much of which involved understanding the terminology and perspectives of the various disciplines, we defined a sustainable agricultural system as one that produces safe and nutritious food at a reasonable price; maintains or enhances the quality of soil, water, and air; minimizes the use of nonrenewable resources; is economically viable; and is socially acceptable. We recognized that these criteria, although not mutually exclusive, were often in conflict and that trade-offs would be required in many instances. The group recognized that a systems approach was needed and proceeded to develop the Guelph FSR program.

Although for various reasons the personnel changed from time to time, the multidisciplinary nature was maintained.

The goal that was established was to develop a system that could be used by individual producers not only to evaluate the sustainability of their system but also, of perhaps even greater importance, to evaluate the impact of policy decisions and/or the introduction of new technologies on the sustainability of the agricultural system in general. Examples of the latter purpose that were suggested included the impact of the milk quota system as it influenced herd size and hence the manure management impacts on the environment; and the impact of the introduction of bovine somatotropin (bST) on the total dairy system.

Realizing that the success of such a program required that there be one person able to devote all or most of his or her effort to the program, a proposal was prepared for funding from the Canadian Tri-Council (Natural Sciences and Engineering Research Council, Social Sciences and Humanities Research Council, and Medical Research Council) for a Chair in Farming Systems Research. This proposal was eventually combined with a proposal for a Chair in Ecosystem Health. The joint proposal was funded beginning in 1994. Since that time, the FSR program has been funded by matching funds from the Ontario Ministry of Agriculture and Food and the Tri-Council Chair in Ecosystem Health as well as other project-related funds.

The FSR team has devoted much effort to developing a framework for problem solving in sustainable farming systems research (Chapter 4) and to describing the boundaries and components of a model farming system, including the linkages among the components. Much effort has also been devoted to the development of indicators of sustainability (Chapter 5). It was recognized early that development of an overall simulation model of such a complex system was unrealistic. Rather, we visualized a series of existing or newly developed simulation models for the different components, the outputs from which could be used to formulate answers to the question being asked.

Not surprisingly, the program has not fully reached its goal. Although time and funding are significant reasons, hindsight suggests that more progress might have been made with a somewhat different approach. I suggest this here not to minimize the progress that has been made but hopefully to provide a guide to future efforts of this nature. We attempted to develop a generalized overall model system in the absence of an identified issue. Although we discussed several possible issues that could be addressed, we did not decide to concentrate on any one. Consequently, our efforts lacked a focus and early deliverables to satisfy the requirements of research managers. As a result, members of the team recognized the need to undertake projects that would provide shorter-term results. A more effective approach would have been to select an issue of current importance around which a model could be developed, while at the same time collecting information of relevance in the short term.

Regardless of these shortcomings, a lot has been accomplished. This book presents much, but not all, of these accomplishments. It is the hope of those involved that it will stimulate others to develop more holistic approaches in their research to ensure that our agricultural systems are sustainable.

Preface

The rise of intensive agriculture and the industrialization of agriculture are part of the latest wave of globalization. The enormous structural changes in agriculture have raised food output, changed the nature of farm structure, and increased the environmental and social consequences of farming. The complexity of these changes can best be understood by an interdisciplinary and coordinated farming systems research (FSR) program in the service of enhanced sustainability within Ontario's increasingly intensive agriculture. This diverges from the dominant production-oriented disciplinary research focus that typifies most Canadian agricultural research. Utilizing FSR conducted mainly in Ontario and, to a lesser extent, Alberta, this book presents an alternative farming systems analysis of environmental, social, and economic factors associated with intensive farming systems.

At the outset, readers are reminded of the devastating consequences of the willingness of some people to ignore the full costs of guaranteeing the public's access to a dependable supply of safe, potable water. The fact that in May 2000 over 2,000 people were hurt and seven killed in Walkerton, Ontario, by untreated water containing *Escherichia coli* bacteria from a cattle farming operation near a faulty well has drawn attention to the need for maximum diligence in handling livestock. But many other social and economic issues are also raised by the changing nature of Canadian farming. The intensification of agriculture has provided an abundance of cheap, largely nutritious, and safe food demanded by cost-conscious citizens who reside mainly in urban areas. It has, however, also been accompanied by controversy over the use of hormones in beef, antibiotics in pork, excessive phosphorus and nitrogen in water, soil erosion, greenhouse gas production, and reduced biodiversity. Added to the mix are issues created by the growing nonfarm rural population, which has often generated conflict between farmers and exurbanites living in rural areas.

In the wake of the Walkerton tragedy, the more numerous nonfarm and urban populations of Ontario have sought to impose environmental regulations that farmers had earlier been hoping to avoid through voluntary compliance. The risk of contamination due to inadequate water treatment and manure management in Ontario culminated in the passing of the Nutrient Management Act in June 2002.

After describing some of the major environmental and social problems connected with Ontario's most productive intensive farming region, this book outlines a framework for analyzing the sustainability of these farms. While the farming systems research applications that follow the framework vary from the relatively more interdisciplinary to "coordinated disciplinary" farming systems research, they all address questions arising from environmental, social, and economic problems of these farming systems.

The book is divided into four parts. Part 1 provides an overview and looks at the issues. Chapter 1 describes how changing farm structures have evolved in the context of globalization and the internationalization of food regimes. It then explains why an interdisciplinary FSR approach is suited to the analysis of these changing conditions. An overview of some of the main environmental and social problems accompanying the growing concentration, centralization, and specialization of these farming systems follows in Chapters 2 and 3.

Part 2 of the book discusses a method for understanding the key linkages among the environmental, economic, and social indicators. Some of the ways in which this framework can be applied in order to understand socioeconomic and environmental linkages within Ontario's intensive farming systems are then presented. Methods for modelling the bigger issues that are broader than farm productivity and viability but that impact environmental sustainability are outlined at the end of Part 2. Despite the fact that the modelling components elaborated arise specifically out of the agricultural economics discipline, an effort is made to integrate this "coordinated disciplinary research" with the overall farming systems analysis of sustainability.

Part 2 begins with a description of how farming systems research at the University of Guelph originally developed and became focused on intensive agricultural systems in southwestern Ontario. It presents a problem-solving framework that can be used to research and solve agricultural problems of an environmental and socio-economic nature. It contains three methodological chapters that establish the systems framework for the FSR research; describe a method of linking the various social, economic, and environmental indicators; and then explain how farming system linkages can be modelled.

Chapter 4 reviews the sustainability problem-solving framework being used by the FSR team at the University of Guelph, which has addressed important issues in both farming systems analysis and sustainable agriculture. The framework, which is grounded on the premise that North American farm operators will be able to adopt more sustainable practices only when they have clear standards against which to gauge their success or failure, has been presented as an integrated and practical method focused on solving problems in sustainable agriculture.

The framework is integrated in several ways. First, and in the most general terms, it draws together traditional farming systems methods and sustainable agriculture issues. However, it acknowledges that, when applied to North American agriculture, FSR must also be cognizant of global competition and regulatory and consumer demand pressures. The framework deals with farm-level applied problems using a collaborative, conceptual, analytical, and evaluative approach. Once a specific problem has been articulated, the system functioning can be analyzed and results implemented in improved techniques.

A second integration feature involves the multidisciplinary approach that the FSR framework embodies. Although formal scientific analyses may be confined to disciplinary fields, they become synthesized when solutions are generated in a later phase and in the context of a primary goal. This process requires integration across disciplines and between the realms of public interest and scientific expertise. In the FSR framework, neither the realms of public interest or scientific expertise can operate effectively without the other. Equally possible, and highly desirable, is the development of more formally integrated models that capture the linkages and interactions between environmental, economic, and social dimensions of farming systems and facilitate whole system-level prediction. The development of such tools is a needed component and a valuable by-product of systems evaluation research.

Sustainability evolves as the socio-cultural context changes. The research framework described in Chapter 4 provides a structure that accommodates disparate needs, problems, and solutions. It is presented as a practical approach in applied farming systems research to support sustainable agricultural systems. Our choice of predictive and summative indicators that are biophysical, social, and economic provides avenues for data collection and analysis from the various disciplinary perspectives included in the FSR team. Our FSR work incorporates both interdisciplinary and coordinated disciplinary work based on agricultural economics modelling of trade-offs between profitability and environmental costs, among various sustainability options that farmers face.

The integration of socio-economic and biophysical variables is complicated in part because the manner in which some of these variables interact at different spatial and temporal scales is often not well understood. Farming systems research is most challenging when it attempts to understand interrelations of socio-economic and biophysical variables at different levels of aggregation. Concerns about social and environmental aspects of sustainable development add a temporal dimension to the problems being investigated. Such concerns make it necessary to undertake both longitudinal and comparative studies of farming systems.

Chapter 5's distinction between predictive and summative indicators is presented as a tool for understanding the linkages between the farm and the community. In order to develop a sustainable agriculture, indicators can be used as criteria for sustainable production. These descriptive and diagnostic indicators can be designed to take account of threshold-level boundaries that consider problems of scale. The choice of indicators should be consistent with the goals of sustainability at farm, community, and environmental levels. They should also be based on the objectives of particular farming systems. Indicators must be selected so that they will have meaning and credibility to the people most involved with these systems: researchers, the public, government officials, and farmers. The chapter proposes a model of the main components of a farming system that integrates the indicators of sustainability and shows its links to the rural community. To be predictive, farming system models must integrate the indicators of sustainability such as viability, productivity, environmental protection, and social acceptability in such a way that they simultaneously consider social, economic, and biophysical indicators, while taking into account a farm family decision-making component.

Chapter 6 argues that the cause-and-effect relationship between agricultural production systems and environmental health can be incorporated into any analysis designed to provide input on sustainability issues surrounding agriculture. This chapter provides suggestions for researchers attempting to assess aspects of the sustainability of alternative agricultural production systems. Trade-off curves represent a convenient means of summarizing the information for policymakers and form the basis for conceptualizing and empirically modelling issues regarding sustainability.

The need to ensure the consistency of data among the disciplines involved is a major issue facing modellers of agricultural sustainability. The unit of analysis both at the individual decision maker and aggregate policy levels should be defined on the basis of the important economic issues regarding sustainability. The biophysical models that are used to estimate the effect of agricultural practices on resource quality must be able to account for the intensive and extensive management choices that are felt to be contributing to the problem of concern. Chapter 6 argues that the individual unit of analysis tends to be determined by the most appropriate biophysical model available. The researchers must then construct a decision-making model for this individual unit, and in the process consider many of the issues that are common to farm-level models designed to enhance individual returns, such as time and risk. However, the modeller must also consider how to aggregate the results across heterogeneous units to the level at which policy decisions are made, as well as how to handle multiple objectives.

Part 3 provides applications of the FSR concepts and methods for providing solutions to economic and environmental problems of intensive agriculture. These chapters present a combination of "coordinated disciplinary research" and interdisciplinary research. Because much of the debate surrounding the intensification of agriculture centres on the growth of intensive livestock operations (ILOs), the applications begin by looking at the trade-offs between economic and environmental goals. The returns to scale associated with pork production combined with the competitive pressures that have impacted producers in the aftermath of the signing of free trade agreements have encouraged many pig farmers to expand their operations.

Chapter 7 presents a study of livestock manure systems for swine-finishing enterprises that is particularly relevant to intensive livestock operations. This analysis clearly shows that, in the absence of government subsidies to encourage environmentally friendly practices, hog farmers are faced with management decisions that pit economic goals against environmental protection. Competitive pressures have often forced swine producers to put economic goals ahead of environmental goals. This chapter develops a model that shows how it is possible for swine producers to maximize their net returns while being environmentally sustainable.

Chapter 8 reviews the development and use of a computerized decision support system for manure management designed by an interdisciplinary team headed by an agricultural engineer. The expert system program that the authors have developed helps farmers manage their manure in order to reduce cost, labour, and odour; protect nutrient availability; and reduce environmental risk. The chapter describes the main functions of the MCLONE4 program and provides an example of its prediction of relationships between various manure application goals and economic benefits. By detailing how MCLONE4 can be used with a dairy operation, this chapter shows how the program can be used to minimize environmental risks.

Chapter 9 presents the results of a study of the properties of sustainability of Grand River dairy farming systems in Ontario, especially with respect to viability, profitability, and acceptability. The Grand River dairy farmers' excellent perceived quality of life is related to the existence of their orderly marketing system (which tames the chaos of dairying), a strong sense of community and personal relations supporting their spiritual well-being, good regular incomes, and the freedom and independence that go with being their own bosses.

Chapter 10 diverges from the other chapters, which focus on Ontario, by addressing the situation in the Crowfoot Creek watershed in Alberta. It looks at a relationship between farm income and the environment in a situation where a proactive attempt has been made to improve water quality through the use of best management practices. Once again, there are trade-offs between the establishment of riparian buffer strips, for example, and farm income.

The final application, Chapter 11, takes a close look at the status of the Environmental Farm Plan (EFP) in Ontario. This plan, based on the "Farm a syst" program in Wisconsin, was developed as a partnership between the Ontario Ministry for Agriculture, Food and Rural Affairs (OMAFRA) and members of the Ontario Farm Environmental Committee (OFEC). The chapter reviews the changing role and status of the EFP.

The final chapter, in Part 4, summarizes the main lessons learned from the experience of the University of Guelph's Farming Systems Research groups studying Canadian intensive agriculture over the past decade. On the one hand, these lessons concern the linkages and partial decoupling among these increasingly intensive farming systems, their biophysical environments, and neighbouring communities, and what to do about these evolving changes in the light of ongoing globalization and industrialization. On the other hand, these lessons point to the strengths and limitations of interdisciplinary and coordinated disciplinary farming systems research for grappling with the sustainability of these complex systems.

Acknowledgments

The farming systems research (FSR) project began under the direction of University of Guelph Land Resource Science Professor Murray Miller, who enlisted the able help of Dr. Stephan Weise. Support from the Canadian Tri-Council (Natural Sciences and Engineering Research Council, Social Sciences and Humanities Research Council, and Medical Research Council) was made available for the Junior Chair in Farming Systems held by Professor John Smithers (1994-99). Matching funds were provided by the Dean of the Ontario Agricultural College. Support funds to other members of the FSR group were provided by the Ontario Ministry of Agriculture and Food (OMAF).

Directors of the FSR team included Land Resource Science Professor Murray Miller (1992-95), Animal Genetics Professor John Gibson (1995-96), Crop Science Professor Clarence Swanton (1996-2000), and Rural Extension Professor Glen Filson (2000 to present). Sociologist Dr. Ellen Wall acted as FSR Coordinator from 1995 to 2000.

Our website has been continuously maintained by Agricultural Engineering Professor John Ogilvie. Although it has not been possible to include everyone's written work in this book, I would personally like to thank the members of the Social Indicators team, landscape architects Professors Cecelia Paine and Jim Taylor, and agricultural economist Wayne Pfeiffer. Postdoctoral fellows who played a significant role in our research and the preparation of this book include land resource scientists Dr. Dean Barry and Dr. Chris Duke, and ecologist Dr. Svenja Belaoussoff. Many graduate students have had important roles in conducting the research, including Kevin Ma, Susan Mulley, Dr. Kofi Anani, Dr. Nasser Yazdani, Dr. Ellen Klupfel, and Santiago Olmos. Pam Lamba deserves thanks for producing the Glossary and Acronyms sections.

Throughout various research projects, we also obtained valuable financial support from Ontario Pork, the Canadian Farm Business Management Council, and the Henry Schapper Fellowship in Agricultural and Resource

Economics, University of Western Australia. We were also fortunate to have the cooperation and support of the Ontario Federation of Agriculture, the Ontario Environmental Farm Coalition (chaired by Professor John FitzGibbon), the Dairy Farmers of Ontario, the Christian Farmers Federation of Ontario, the Grand River Conservation Authority, OMAF personnel such as Greg de Vos, many individual farmers in Ontario and Alberta, and the North American Chapter of the International Farming Systems Association.

This book has benefited enormously from the advice of individuals who reviewed it for UBC Press and the Aid to Scholarly Publications Programme.

Abbreviations

2,4-D	dichlorophenoxyacetic acid
AAFC	Agriculture and Agri-Food Canada
AAFRD	Alberta Agriculture, Food, and Rural Development
AESA	Alberta Environmentally Sustainable Agriculture
AGNPS	Agriculture Non-Point Source Pollution model
BMP(s)	best management practice(s)
CCWG	Crowfoot Creek Watershed Group
CFFO	Christian Farmers Federation of Ontario
CRP	Conservation Riparian Program
CSA	community shared agriculture
DAFOSYM	Dairy Forage System Model
DEMP	Density Equalized Map Projection
DFO	Dairy Farmers of Ontario
Dicamba	3,6-dichloro-2-methoxybenzoic acid
DSS	decision support system
EFP	Environmental Farm Plan
EU	European Union
FSR	farming systems research
FSR/E	farming systems research and extension
GATT	General Agreement on Tariffs and Trade
GHG	greenhouse gases
GIS	Geographic Information System
GMOs	genetically modified organisms
GP	goal programming
HACCP	Hazard Analysis Critical Control Point
ICR	interactive conflict resolution
ILO(s)	intensive livestock operation(s)
IPC	Integrated Pest Control
IPM	Integrated Pest Management

ISO	International Organization for Standardization
K	potassium
K_2O	potassium oxide
LEACHP	Leaching, Estimation and Chemistry, Pesticide model
LISA	low-input and sustainable agriculture
LP	linear programming
MCLONE	Manure, Cost, Labour, Odour, Nutrient Availability, Environmental Risk
MIP	mixed integer programming
MOTAD	Minimization of Total Absolute Deviations
MP	mathematical programming
MSD(s)	minimum separation distance(s)
N	nitrogen
NAFTA	North American Free Trade Agreement
NGO	nongovernmental organization
NH_3	anhydrous ammonia
NH_3	ammonia
NMA	Nutrient Management Act
NMAN	Nutrient Management Program
NMP	nutrient management plan
NOLP	nearly optimal linear programming
ODFAP	Ontario Dairy Farm Accounting Project
OFA	Ontario Federation of Agriculture
OFEC	Ontario Farm Environment Coalition
OMAF	Ontario Ministry of Agriculture and Food
OMAFRA	Ontario Ministry of Agriculture, Food and Rural Affairs
OMMB	Ontario Milk Marketing Board
P	phosphorus
P_2O_5	phosphorus oxide
PFRA	Prairie Farm Rehabilitation Administration
pH	potential of hydrogen
rbST	recombinant bovine somatotropin
RISE	Response-Inducing Sustainability Evaluation
RWQP	Rural Water Quality Program
WTO	World Trade Organization
ZECs	zones of ecological compensation

Part 1:
Issues and Overview

1
Introduction
Glen C. Filson

For over a decade, the University of Guelph's farming systems research (FSR) team has been conducting interdisciplinary research into the extent to which our more intensive farming systems are sustainable with respect to several dimensions. This book elaborates on our FSR framework with respect to sustainability indicators and assesses trade-offs between the economic, environmental, and social consequences of intensive Canadian agriculture, especially within Ontario. To understand the importance and approach of the research, this chapter explains some of the issues raised by the rise of intensive agriculture on a global scale and highlights some of the environmental and social consequences that are then discussed in Chapters 2 and 3.

The Rise of Intensive Agriculture

The industrialization of agriculture spurred on by international free trade agreements and economic globalization has been occurring as a new food regime has evolved in the past generation. The consolidated corporate control of food chains has led to backward linkages to farms and has facilitated greater concentration and specialization of food production. Huge increases in production in a world with a rapidly expanding population hold tremendous potential to increase people's food security and reduce poverty, if only this production could be equitably shared by everyone. Unfortunately, the benefits from this dramatic increase in agricultural production have been available only to those who can afford to purchase the produce. Considerable voluntary efforts have been made to introduce the forms of best management practices to minimize negative environmental impacts from this food production (see Chapter 11). Nonetheless, in Ontario, public impatience with the speed and comprehensiveness of these environmental efforts have culminated in provincial regulation of intensive agriculture, trumping the myriad of municipal bylaws that have recently been enacted to limit the expansion of some forms of agriculture, particularly animal agriculture.

Traditionally, farming has been the only major industry in which families comprised the largest share of the labour force. Now, as the growth of large, highly mechanized corporate farms continues to increase, especially in Ontario and Alberta, the number of farmworkers is approaching, and in many places exceeding, the number of family workers.

Even though there has been a shift towards more specialized crop production of a greater variety than before, there is a growing monocultural concentration on a few main crops such as soybeans and corn. Increasingly, plants and to some extent animals are being genetically altered to improve their ability to make use of nutrients and grow in specialized conditions (Chopra et al. 1999; Ellstrand 2003).[1] There is also growing evidence that genetically modified foods are further intensifying agricultural production, leading to greater economies of scale and concentrating production into fewer, larger farm units.[2]

Control of the Canadian food system resides mainly in the hands of large corporations, often to the disadvantage of smaller agribusinesses and farm operators as well as consumers. Winson's rural sociological study of the agro-industrial food complex in Canada concludes that "there is no doubt that in the food system, at least, we are witnessing the accelerated polarization of resources, both within the farming community and among the various players in the agro-food complex" (1992, 210).

The latest "food regime" (comprising many food chains from conventional agricultural production and distribution to organic) is strongly affected by greater globalization, liberalization of trade, and a wave of privatization as once perishable food is now transported throughout the world. The most recent food regime has seen increased market competition, lower food prices, and a consolidation of corporate control of food chains favouring bigger, more industrialized farms. The food system extends outward from farms towards institutions that include farmers, food processors and wholesalers, food retailers, government policymakers, and consumers (Winson 1992). The effects of the latest food regime are also visible within rural communities as conflicts between farmers and nonfarmers have often worsened in many countries (Friedmann and McMichael 1989). Friedmann believes that, just as in the post–Second World War period, the food system remains in crisis, but now it coincides with a growing environmental dilemma. Whereas the total quantity of food was the primary postwar concern, now food quality, especially food safety, has become a major concern, along with the ecology of agriculture (Friedmann 2002).

---

1 U.S. Department of Energy, "Genetically modified foods and organisms," online at <http://www.ornl.gov/sci/techresources/Human_Genome/elsi/gmfood.shtml> (retrieved 21 November 2003).

2 Economic and Social Research Council (ESRC), Global Environmental Change Programme, "The politics of GM food: Risk, science, and public trust," online at <http://www.sussex.ac.uk/Units/gec/gecko/gm-brief.htm> (retrieved 28 November 2003).

Globalization is "the accelerated integration of capital, production, and markets driven by the logic of corporate profitability" (Bello 2003, 1). Trade liberalization, privatization, and deregulation are additional features of globalization. The first phase of globalization, which occurred from the nineteenth century to the end of the First World War, was ended by the rise of national capitalist economies that had significant state intervention and limits on capital flows and trade. The second phase, Bello argues, began with the structural adjustment programs of the 1980s followed by the establishment of the World Trade Organization (WTO) in 1995. It continues to the present as the International Monetary Fund, the World Bank, and the WTO work to create economic models of international governance.

The effect of globalization and accompanying freer trade on Canada's farmers has been to force them into greater technological mechanization and industrialization of their farming operations, as they must now compete not only with their neighbours but with the more industrialized and subsidized farming operations that increasingly exist in other parts of the developed capitalist world. As shown in Chapters 2 and 3, the growth of these more intensive farms can have serious negative impacts on the environment and the rural communities that are affected by changing farm structures.

While farming is essential to most human survival, nonpoint source water pollution from animal waste, chemical fertilizers, antibiotics, and pesticides threatens water quality in the developing and developed world. Intensive agriculture can also produce excessive amounts of greenhouse gases such as methane and carbon dioxide, contributing, in turn, to global climate change (see, for example, Boyd 2003). Despite these and other problems, such as accompanying soil compaction and erosion (see Chapter 2), however, larger, more industrialized farming operations are probably here to stay and they are not necessarily any harder on the environment than a myriad of small farms. It is therefore incumbent upon the agricultural research community to describe and analyze these operations so that the most amenable social, economic, and environmental accommodations can be made for the greater good of all food producers and consumers.

For our purposes in this book, "intensive agriculture" is defined as farming operations practised under conditions where there is an increasing tendency for many farms to have become relatively large despite the fact that many are still family-operated farms. The larger operations[3] rely increasingly on a

3 D. Galt and T. Barrett, Task Force on Intensive Agricultural Operations in Rural Ontario Consultation, "Summary of consultations," online at <http://www.gov.on.ca/OMAFRA/english/infores/releases/081603_b.html> (retrieved 6 June 2002). *For livestock farms:* Many suggested that a livestock farm with more than 150 livestock units (LU) on a farmstead site is intensive. Others suggested there be at least 600 livestock units. And others suggested 1,500 livestock units. Several people felt that the density of the number of animals

combination of mechanized forms of production, the use of fertilizers and other agri-chemicals such as pesticides and herbicides, increasing use of biotechnology, and a small but growing number of agricultural workers. They therefore employ relatively larger investments in land, labour, and capital than was traditionally the case when smaller, more mixed farming operations predominated. Thus, even though many small, medium, and large farms still function within southwestern Ontario, particularly in some branches of agricultural production, the larger farms continue to grow, displacing many smaller farms.

Tables 3.1 and 3.2 show how substantial increases in numbers of animal units per farm in Ontario have been. Thus, while intensification also affects cropping systems, there has been significant growth in the concentration of pigs and chickens in particular, especially in the southwestern part of Ontario. This can also be clearly seen in the maps in Chapter 11.

In recent years, a number of southwestern Ontario citizens' groups have been formed in Middlesex, Bruce, and Peterborough counties with the express purpose of limiting the expansion of industrial agriculture and manure production. Many townships and counties have passed bylaws placing moratoriums on the expansion of large livestock operations and requiring nutrient management planning.[4]

Intensive livestock operations have often raised the public's hackles most. Ontario Environment Commissioner Gord Miller has pointed to concern that many of the new large livestock farms producing huge amounts of manure do not have equally large areas of farmland to accommodate all of the manure generated.[5] He observed that Ontario's pigs now produce as much raw sewage as the more than 10 million Ontario people. Ontario's cattle, of course, produce substantially more manure than all of the pigs and humans combined.

Christian Farmers Federation of Ontario (CFFO) strategic policy adviser Elbert van Donkersgoed has also asked whether "agricultural intensification, driven primarily by productivity increases and economic efficiencies,

---

should be used as the criterion. Examples included greater than 1 LU per acre; greater than 1.5 LU per acre; and greater than 2 LU per acre. *For cash crop farms:* Some felt that farms having more than 2,000 acres should be considered intensive; others felt the number should be more than 5,000 acres. *For greenhouse operations:* Any operation having more than 15 acres under glass was suggested as being intensive.

4 Ontario Ministry of Agriculture and Food (OMAF), "Municipal nutrient management plan review," online at <http://www.gov.on.ca/OMAFRA/english/nm/nman/municipal.htm> (retrieved 6 April 2004).

5 G. Miller, "The protection of Ontario's groundwater and intensive farming." Special report to the Legislative Assembly of Ontario by the Environment Commissioner of Ontario, 2000, online at <http://www.eco.on.ca/english/publicat/sp03e.pdf> (retrieved 2 March 2001).

[has] come so far, so fast, that it cannot continue without creating unacceptable consequences for our environment and rural communities?"[6]

Walkerton's Contaminated Water Supply Forces the Government to Regulate Intensive Agriculture

In Walkerton, a town of 5,000 in southwestern Ontario, over 2,300 people were injured and 7 were killed in May 2000 by a relatively new strain of *Escherichia coli* bacterium arising from a local beef operation. Tragically incompetent water treatment procedures, poor communication between privatized laboratories and the local board of health, and a government bent on saving money by "managing the risks" created by a severely downsized Ministry of the Environment all conspired to put intensive agriculture in the public spotlight.

Mayor David Thomson admitted to having been "dumbfounded as to what had been dropped on us." The deadly strain of *E. coli* O157:H7 was soon traced to a herd of Limousin cattle owned by an equine veterinarian who had completed an Environmental Farm Plan (EFP) for his 565-hectare, mainly cash-crop farm (Bourette 2000). Just before the crisis, the Ontario Ministry of Agriculture, Food and Rural Affairs (OMAFRA) "had been crisscrossing the province trying to prevent municipalities from issuing local bylaws to stop large-scale factory farms" (Gallon 2000). OMAFRA[7] officials had been anxious "to protect intensive industrial-livestock operators from challenges by local residents worried about odour and groundwater contamination" (Mittelstaedt 2000). Despite the fact that the Walkerton tragedy could not be blamed entirely on Walkerton's Public Utilities Commission, agricultural production, or the Ontario government, one of its main effects has been to sharpen concern about the environmental effects of intensive farming.

Since the Walkerton tragedy, the relationship of intensive agriculture to the environment and people's well-being has been a public issue for Canadians. The second part of Justice D. O'Connor's *Report of the Walkerton Inquiry* (2002) recommended that sensitive-area small farms and all large farms be required to develop water protection plans, although the government has yet to require this. Justice O'Connor also recommended that while the Ministry of the Environment should regulate the impact of farming practices on drinking water sources, he recommended that "all large or intensive farms, and all farms in areas designated as sensitive or high-risk by the

6 E. Van Donkersgoed, "Concern about intensive agriculture is about more than manure," *Corner Post*, 1 July 2000, online at <http://www.christianfarmers.org/commentary/cpost/2000/cp-01-07-00.htm> (retrieved 14 August 2001).

7 The Ministry has since given up Rural Affairs and is now once again simply called the Ontario Ministry of Agriculture and Food (OMAF).

applicable source protection plan, should be required to develop binding individual water protection plans consistent with the source protection plan" (O'Connor 2002, 20). In the aftermath of the Walkerton tragedy, this initiative has now achieved provincial regulatory backing with the passage of the Nutrient Management Act (NMA) in June 2002.

Ontario's NMA establishes categories according to farm size and practices, so that specific standards can be created for each category and compliance required by municipalities. Also included in the legislation are specified minimum separation distances (MSDs) between buildings, manure storage facilities, and manure applications, as well as between such facilities and watercourses, sloughs, and swamps. Nutrient management plans (NMPs), voluntary since 1997, may eventually be required for all farmers, as originally envisaged in the act.

Since the passage of the legislation, criticisms of the NMA by Ontario's major farm organizations, including the Ontario Federation of Agriculture (OFA), the CFFO, and the Ontario Farm Environmental Coalition (representing those organizations and another twenty-eight farm organizations) have delayed the implementation of the act, especially for small farm operations. Views about nutrient planning among farm organizations on the one hand, and environmental and municipal groups on the other, differ and this has created considerable conflict in many rural areas (see Chapter 3). Given the complexity of modern farming and the relationship it has with the rural and urban communities with which it is articulated, we believe that an analysis that takes a systems approach and is genuinely interdisciplinary provides considerable analytical power for confronting the most salient issues.

Farming Systems Analysis of Intensive Agriculture

Using a systems approach, members of the FSR group have studied the properties and dimensions of the sustainability of southwestern Ontario farming systems in order to provide solutions to air, water, and soil quality problems as well as such interactional difficulties between farming and nonfarming communities as the lack of understanding of modern farming. Work of this nature is an effort at re-establishing a farming systems discourse based on a synthesis of existing disciplines.

When the term "farming systems research" is invoked, the whole farming system is being considered instead of such isolated aspects of a farming system as its crops, livestock, or technologies. FSR is also concerned with farming households, their profitability, and their social and environmental impacts. Issues affecting the farming systems beyond the farm gate, including the political economic climate within which those farms function, their inputs and outputs, and their socio-cultural environment, are also of concern.

Systems analysis is a useful mechanism for understanding and integrating the relationships of biophysical and socio-economic disciplines among different spheres by viewing farming systems as arrangements of component parts that continuously interact to achieve goals by transforming inputs into outputs. Ecosystems such as the Grand River Watershed, within which most of the farming systems we are studying exist, are part of integrated wholes whose properties have to do with the relationships between their components. These systems extend outward and are also part of the international food system. Each subsystem is profoundly affected by the biophysical realm, which in turn affects society, all of which is affected by international food systems although each system and subsystem has its own specificity.

The systems approach is not only very useful for understanding how sustainable our farming systems are but also allows us to consider the farming systems in the context of the ecosystems within which they are situated and the rural communities of which they are component parts. Seeing these farming systems as part of a global food regime or system helps make the linkages between these rural communities and dominant socio-economic and environmental systems. The systems approach also connects researchers with farmers, policymakers, and the consumers of food and fibre.

Systems analysis is an integrative way of applying the scientific method across a variety of scales and disciplines. Although more difficult, modelling complex economic and ecological systems as a group instead of separately enables assessments of the ecological and socio-economic interactions to be taken into account. These intricate systems have strong interactions among their components and complex feedback loops, making the cause-and-effect relationships among parts difficult to grasp. An understanding of these interactions is required in order to identify and support the most sustainable activities within both the ecological and socio-economic realms (Costanza et al. 1993).

With complex systems, when a critical mass of the elements of a trend at a lower level of hierarchy within the biosphere occur and affect a higher level's slower process, feedback can have a potentially dramatic effect. Costanza and colleagues (1993, 549) provide the following example: "The rapid and extensive human uses of fossil fuels could be seen as such a trend, causing perturbations at the global atmospheric level, which might feed back and radically alter the framework of action at the lower level." Thus, while the individual actions of fossil fuel consumption may not immediately appear to result in climate change for the world as a whole, complex systems theory has been able to show that rapid fossil fuel consumption has been occurring that is altering both what can be grown in specific locales (requiring adaptation) and what ought to be grown (to partially mitigate climate change) (CCIAD 2002).

Applying the systems concept to a farm, Dent et al. (1995) have set out the minimum conditions required to develop a whole-farm model. These would include farm family and farm components such as type of farm enterprise, management, ownership, off-farm work, the farm family's basic survival needs, motivation, and objectives. These components in turn are affected by economic and environmental forces and the factors that influence the sociocultural characteristics of the farm family, including their education, kinship patterns, access to information, and so on. This is a daunting task because of the level of complexity associated with human systems, which is far in excess of what we have attempted in this book, especially with respect to the socio-cultural and quality-of-life components. As Barry (1998, 3) observed, were it to be developed "the farm family decision making component would most likely be a rule based expert system connected to biophysical, economic, and social or quality of life modules."

Based on Barry's discussions with Ronald Shaffer at the Center for Community Economic Development at the University of Wisconsin and the awareness that, on average, more than half of Canadian farm family income in the 1990s was from off-farm work, Barry outlines a useful way of utilizing farming systems analysis to show how environmental damage can result from intensification of production practices. Figure 1.1 shows how various small-farm linkages can generate income pressures resulting in greater intensity of farm practices, which may then culminate in more environmental damage. On the one hand, off-farm work enables income to be increased without damaging the environment. More labour is usually required by less intensive production practices but these systems allow less time for those farmers to work off the farm. As Barry notes, "some farming systems may therefore develop to accommodate off-farm work requirements, or may only be possible if income is supplemented by off-farm work" (1998, 14).

Within our FSR projects we have therefore used systems thinking to understand the problems with which we are faced. Systems thinking also makes use of mathematical modelling of variables as an essential tool for understanding environmental and economic trade-offs within intensive agricultural systems. When farming systems goals are complex and at times in conflict, such modelling can provide invaluable decision-making support for farmers seeking, for example, to manage their manure in the most economically beneficial and environmentally friendly manner.

Prior to this, most FSR work has been done with diversified small family-run farms in developing countries and has been oriented towards increasing equitability and efficiency among those farms through interdisciplinary extension work. The approach has been on-site and geared towards improving farmers' management decisions. FSR traditionally worked with "recommendation domains," which are essentially homogeneous types of farms with similar sizes and circumstances. Such farms can employ the uniform

Figure 1.1

Linkages between some factors influencing on- and off-farm income

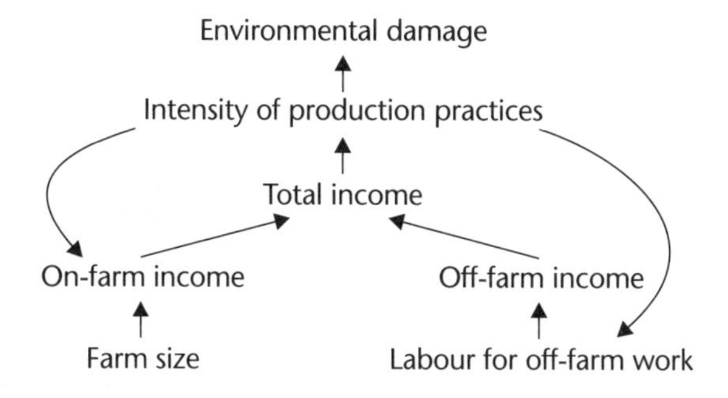

Source: Adapted from Barry 1998, 14 (Figure 9).

types of practices and technologies to enhance productivity in that particular recommendation domain (Hildebrand 1986). The FSR approach is famous for using participatory, bottom-up approaches instead of the more top-down extension approaches associated, for instance, with the World Bank–funded Training and Visit system (Röling 1985; Johnson and Claar 1986; Gibbon 1994). FSR has also typically taken more account of gender relations and household activities (Sellen et al. 1993) than is described here, but our team has discussed Grand River farmers' gender relations elsewhere (Klupfel and Filson 2000). In this respect, I would agree with Norman that while FSR has attempted to incorporate the household in a participatory manner for diagnostic purposes and the design of new technologies, "the assumption of a monolithic household" (2002, 6) does not work and certainly is inappropriate for multicultural Ontario farms.

Employing a holistic perspective that identifies linkages in its integrated description of farm production and consumption, farming systems research also seeks to highlight problems for which it can suggest solutions. FSR thus links the farming system to the rural environment in which it exists. By describing the farming system's household characteristics and the physical, socio-economic, and biological factors under the control of the household and the farm's management, it promotes the farming system's sustainability and the well-being of those within the household and potentially within the farming system's environment. If, for example, there are negative spillover effects on the rural community's environment from the farming system, FSR tries to identify those effects and mitigate them (Sellen et al. 1993).

In Norman's review of the history of the farming systems approach, he acknowledged that "the application of the farming systems approach to livestock enterprises was generally particularly weak" (2002, 6). As this book's focus is more about intensive livestock production than plants, we have tried to overcome this typical FSR weakness.

On the other hand, recent efforts to develop the "farming systems approach with a whole farm focus" (Norman 2002, 6) using tools like Participatory Rural Appraisal, intra-household relationship, and adaptability analysis have largely not been attempted here despite these tools' advantages.

Increasingly, sustainability has been a concern of FSR (Lightfoot and Noble 1993; Lightfoot et al. 1993; Norman 2002). Sustainability properties and dimensions vary widely. Okey (1995), for instance, in assessing the use of these properties and dimensions with agroecosystems, argued that they include resilience, stability, self-organization, efficiency, equity, and diversity/complexity.

The rise of farming systems with a sustainable livelihood (SL) focus based on the work of Chambers (1995), Scoones (1998), and Ellis (2000) has been most useful for looking at the people-centred, holistic, interdisciplinary SL approach that builds on people's strengths and stresses micro/macro linkages. The SL approach to FSR supports the diversification of the livelihood base for farmers into micro-enterprises and cooperatives, which could have applicability in Canada, but at this time its main focus has been on vulnerable farm households experiencing food insecurity in developing countries.

Conway (1987), who has worked on the SL approach with Chambers in developing countries, has defined sustainability as the system's ability to maintain productivity in spite of some major disturbance such as farmers' indebtedness, soil erosion, or drought. Advocating an approach similar to the one taken in this book, he points to four essential properties of agroecosystems: productivity, stability, sustainability, and equitability (1991). Conway considers sustainability to be the ability to maintain productivity despite major disturbing forces affecting the agroecosystem. He recommends trade-offs among these properties of sustainability. For instance, an overly equitable approach could jeopardize productivity, although productivity, which may depend upon the extensive use of chemicals, should probably be sacrificed to some degree to improve sustainability and equitability.

This is, however, a narrow way of viewing sustainability, in Healey's opinion (2003). While she sees Conway's approach to agricultural sustainability as practical and project-oriented, she feels that it misses the crux of the problem, which is that capitalism, which requires growth, necessarily overproduces, thereby threatening the environment as well as the less well off; she therefore feels that this socio-economic system is ultimately unsustainable.[8] Others also say that there is an inherent contradiction built into the

8 As Marx (1959, 813) put it: "Large-scale industry and large-scale mechanised agriculture work together. If originally distinguished by the fact that the former lays waste and destroys principally labour-power, hence the natural force of human beings, whereas the latter more directly exhausts the natural vitality of the soil, they join hands in the

sustainable development notion popularized by the United Nations report *Our Common Future* (World Commission on Environment and Development 1987). O'Connor (1994), for instance, feels that capitalism itself is unsustainable because of its tendency to overproduce relative to the demand for its products within each business cycle. Nevertheless, while the establishment of a more balanced socio-economic and ecological socialism may ultimately yield the support needed to displace industrialized capitalist agriculture, and possibly much sooner in Bolivia, which is of most concern to Healey (2003), we assume that in the near term it is precisely the kind of project-by-project, practical trade-off solutions that will corral the immediate socio-economic and environmental problems that accompany intensive conventional farming in Canada.

What we take to be practical involves what Berkes and colleagues (2003, i) have recommended for a 2003 Canadian Institute for Advanced Research (CIAR) workshop: "researching the lessons to be gained from disturbances in social-ecosystems at several scales, focusing on complexities, thresholds, and interdependencies of these systems; and by integrating social and ecological dimensions in both our research focus and our practices." This, they argue, requires that the hierarchy among disciplines be overcome through humility and hope so that people from the natural sciences, whose quantitatively oriented disciplines often trump those of the social sciences, be prepared to work collegially with social scientists to solve the kinds of "wicked problems" that are normally too complex to be dealt with by systematic, rational processes. Berkes and colleagues (2003) feel that this is particularly important today because, whereas past ecosystem models emphasized the stability of environments using equilibrium models, "our current understanding of ecosystems emphasizes multiple equilibria, surprise, uncertainty, thresholds, and system flips" (Gunderson and Holling 2002).

Features of systems theory discussed within the farming systems analysis presented in this book involve social and biophysical scientists in dialogue about the information flows from models of cropping and animal systems, expert systems to manage manure and control waste, and human resource development models. After the main environmental and social problems confronting intensive agriculture in Ontario are outlined in Chapters 2 and 3, a framework for assessing the sustainability of farming systems is elaborated. Descriptive and diagnostic indicators are discussed as useful methods to be employed, along with predictive and summative indicators to better understand linkages between socioeconomic and biophysical aspects of farming systems. Part 3 illustrates these linkages and the linkages

---

further course of development in that the industrial system in the country-side also enervates the labourers, and industry and commerce on their part supply agriculture with the means for exhausting the soil."

between farming systems and the rural communities within which they reside with studies of dairy, pork, beef, and nonfarm rural communities and the environment within Ontario's Grand River Watershed and Alberta's Crowfoot Creek watershed. There are also applications seeking to balance environmental and economic concerns in manure management, farm-level modelling of economic and environmental issues, livestock manure system analysis for pork, and water quality initiatives.

Farming systems can be extremely complex, so decisions must be made to select the most important components and relationships of particular systems so that models based on minimum data sets can be produced. This selection process requires dialogue and reasoned argument. With various degrees of success, we have developed a dialogue among land resource, crop, and animal scientists, landscape architects, an agricultural engineer, agricultural economists, geographers, a sociologist, and a rural extensionist. All group members have carried out separate but related research so that the work has varied from relatively more integrated interdisciplinary work to less integrated, multidisciplinary work on related problems associated with agricultural systems in southwestern Ontario. The FSR committee itself has continually evolved.

We have spoken with people in farm organizations and on farms to find out what they think about their quality of life, how they manage their manure, whether they feel they can compete in the face of international free trade agreements, and what they think environmental pressures will do to their costs of production. We have also used the conference/workshop approach to engage farmers, academics, and government personnel on farming systems research and on sustainability issues such as the risks posed by climate change.

Many small family farm operations are threatened by the rise of intensive agriculture because they are less capable of making sufficient capital investments to obtain returns to scale and remain competitive in the new low-price international food production market. With the rise of Ontario's increasingly intensive agricultural production at a time when greater numbers of nonfarm urbanites have moved into rural areas, new threats to people's quality of life and the environment have emerged. Pressures to industrialize agricultural production have increased, along with the emergence of a new "food regime" as freer international markets and intensified competitive pressures mark this latest phase of economic globalization.

In the following chapter, the environmental issues associated with agriculture in southwestern Ontario are outlined in greater depth. Some of the related social issues are then discussed in Chapter 3.

2
Environmental Problems Associated with Intensive Agriculture

Glen C. Filson

The pressure to move to a more sustainable agriculture arises from the need to continue to meet economic goals for farm productivity, viability, and social acceptability while limiting pollution and conserving our resources.[1] These sustainability goals have come into sharper focus as globalization trends have crystallized collective efforts to combat climate change through the move to ratify the Kyoto Protocol.[2] This chapter uses environmental indicators to outline both some things that have improved locally, such as soil cover, water, and tillage erosion, and other things that have worsened, such as declining water quality and reduced biodiversity. After reviewing how the Agriculture and Agri-food Canada (AAFC) Environment Bureau has monitored key environmental indicators, this chapter identifies some of the ways in which Ontario farmers are trying to ameliorate the negative environmental impacts of farming as well as react to the criticisms they face over their farming practices.

Environmental Problems Associated with Agriculture

The rise of modern agricultural production systems and their possible threat to the environment have been addressed by a number of national and regional studies. In 2000, AAFC published *Environmental Sustainability of Canadian Agriculture,* a comprehensive national study of agri-environmental indicators for use by farmers and their leaders, environmentalists, policymakers,

1 This chapter benefits from data collected with the help of generous financial support from the Ontario Ministry of Agriculture, Food and Rural Affairs (now simply the Ontario Ministry of Agriculture and Food, or OMAF).

2 This is now under threat of not passing because of the Russians' refusal to ratify it. Although 120 countries have ratified it as of December 2003, the second trigger necessary for it to go into force, "that the ratifying governments must include developed countries representing at least 55 per cent of that group's 1990 carbon-dioxide emissions, remains to be met" (Chase et al. 2003). Without the support of the US and Australia, Russia's ratification is crucially important to move the 44.2% total emissions of ratified countries above 55%.

and the general public. Although complementary to some of the indicators we have used, unlike ours the AAFC indicators were not designed for application at the farm level. So, while AAFC researchers discuss nonpoint sources of pollution, for example, they do not mention the extent to which specific intensive livestock operations contribute to groundwater or surface water nitrogen and phosphorus pollution. They concede that their research results cannot be used as a guide to best management practices on farms despite the fact that environmental farm management is one of their six indicators. They conclude by noting that "this more-intensive form of agriculture in an environment where water supplies are abundant increases the potential for agriculture to have adverse environmental effects" (McRae and Smith 2000, 195). However, MacGregor and McRae (2000) also observe that demand for agricultural products will continue to increase and, as output expands, the risks to the environment will also increase.

World market–oriented, profit-maximizing conventional agriculture, which has become increasingly intensified in Canada, not only has led to overproduction of farm produce but has also had pronounced effects on biodiversity, deteriorating parts of the environment in the process. Häni (1998, 2001) argues that growing recognition of this consequence first generated a combination of Integrated Pest Management (IPM) and Integrated Pest Control (IPC) as a way of minimizing the damages caused by conventional agriculture. While these solutions had some beneficial effects, Häni describes IPM and IPC as part of a reductionistic strategy because of the primary limitations that they did not adequately address the main causes of pest problems. These causes, he argues, are excessive inputs of N-fertilizer, inadequate crop rotation, and the use of varieties that are susceptible to pests. Second, Häni observes that IPM focuses only on individual pests. This makes it difficult "to simultaneously use many different specific control strategies for individual pests, diseases, and weeds of several crops" (2001, 1).

Despite the rise of intensive agriculture, it is important to observe that alternative forms of agriculture have also grown over the past decades. The worldwide tendency for subsidies to be reduced due to free trade has led some farmers to intensify their operations, while others are switching from the conventional, high-input strategy to more sustainable practices, the only ones, for instance, that are subsidized by the Swiss government. The latter include greater use of natural regulation and other farm resources (Häni 2001).

Holistic agriculture developed many converts in the 1980s in both Europe and North America. Variations have included low-input and sustainable agriculture (LISA), ecological, alternative, and organic agriculture, all of which share an ecosystem-oriented instead of a world market–oriented vision. While organic agriculture is the most radical version of this vision, eschewing any use of synthetic pesticides, integrated farming systems try to maximize the use of natural regulation while minimizing the use of pesticides (Häni 2001).

Despite the rise of these alternative, holistic forms of agriculture, especially in such European countries as Switzerland and Sweden, Statistics Canada's 2001 Agricultural Census revealed that the numbers of these integrated farming systems was still quite small in Canada. Still, their rate of growth is quite high (Gillespie 2001).

Returning to the AAFC's agricultural indicator approach, the central conceptual framework used is based on the notion of Driving Force–Outcome–Response. "Driving force" refers to the forces influencing agricultural activities. The outcomes are the environmental consequences of agriculture. "Response" refers mainly to societal responses to any changes that occur in the driving forces or the outcomes with respect to what technologies farmers adopt, how consumers react, and the reactions of governments. Although beyond the AAFC's mandate, the reports, unfortunately, say little about rural community/agricultural interactions, conflicts, and changes in sense of well-being.

The AAFC scientists and academics commissioned by AAFC studied water and soil quality, greenhouse gas emissions and climate change, production intensity, and the impact on biodiversity from agroecosystems.[3] Despite the absence of micro farm and rural community level analysis, their research is the most comprehensive study of agricultural indicators ever undertaken in Canada. They provide us with useful baseline data at regional and national levels that will be of enormous benefit to farmers, policymakers, and academics in understanding the linkages between the environment and agriculture.

Häni (2001) uses a similar approach in his Response-Inducing Sustainability Evaluation (RISE) except that it is geared to the micro, individual farm level, but his approach does take the social aspects into consideration. Häni (2002, 1) notes:

> RISE is based on twelve indicators for the economic, ecological and social situation: Energy consumption, water consumption, situation of the soil, biodiversity, emission potential, plant protection, wastes and residues, cash flow, farm income, investments, local economy, social situation of farmer family and employees. For each indicator the "Driving force" (D) and the "State" (S) are assessed.

Häni has been doing this in Switzerland, China, and Brazil with individual farms, identifying how the driving forces and states of these twelve indicators function in the farm's actual situation in order to then optimize the sustainability of the farm by, for example, improving its manure management.

---

3 An extremely useful addition to the AAFC Environmental Bureau's work came out in 2000, entitled *The health of our water: Toward sustainable agriculture in Canada* (Coote and Gregorich 2000).

Häni observes that finding useful indicators of ecosystem health is not easy. Pesticide and fertilizer residues can be found within the air, land, and water. Bioindicators of environmental sustainability include the presence of earthworms, Collembola, mites, and other indicator animals and plants. These are particularly useful at the micro level of farm analysis (Häni 2001).

The questions of how to bring about environmental changes in an economical fashion as well as whether, and to what extent, regulations will be needed to curb the most undesirable aspects of more intensive agriculture are central to the work of AAFC's Environment Bureau. While they look closely at the environmental implications of the growing shift to more intensive agriculture, AAFC scientists point out that these structural changes have increased pressure on environmental resources at the same time that environmental expectations are rapidly evolving.

Our FSR studies instead focus more specifically on the most intensive agricultural region of Canada, southwestern Ontario, and to a lesser extent southern Alberta. However, the AAFC research provides the macro picture within which our micro work can best be understood. While some of the context for the study on farming in Alberta is provided in the chapter on the Crowfoot Creek Watershed (Chapter 10), the environmental review of agriculture's effects on the environment will be largely restricted to Ontario, especially in the southwest, where most agriculture is concentrated. As will be seen, the relationship between growing intensification of agriculture and competing and growing urban incursion into the area has heightened conflict, especially between livestock production, the environment, and competing social demands on the resources of the area.

Before looking specifically at some environmental costs and benefits of Ontario farming, the major aspects of the ecosystems in Ontario and specifically the southwestern region should be briefly summarized. About one-quarter of Canada's primary agricultural Gross Domestic Product is accounted for by Ontario, and a substantial part of that is contributed by southwestern Ontario.

The agroecosystem of southwestern Ontario is Mixedwood Plains with "gentle topography, fertile soils, a warm growing season, and abundant rainfall" of 1,000 millimetres of precipitation annually (McRae and Smith 2000, 188). The risk of water contamination by nitrogen increased from 1981 to 1996, although the water remains near or above drinking water standard in most areas. Also, 37% of the farmland assessed by AAFC had over 60 kg of nitrogen per hectare, an increase of about 5 kg/ha from 1981 to 1996. The rise in residual nitrogen levels was due mainly to increasingly intensive livestock production and the increased area of land that is under crops requiring high nitrogen inputs, such as corn (McRae and Smith 2000).

On the positive side, indicators for Ontario of soil cover, water and tillage erosion, and soil carbon all improved from 1981 to 1996, while only soil compaction worsened. Much of southern Ontario has been significantly compacted as the soils susceptible to compaction grew by 61% between 1981 and 1996 in Ontario (McBride et al. 2000). Water quality also deteriorated, mainly through increased bacterial contamination. According to Fairchild and colleagues (2000, 61):

> The incidence of bacteria in well water appears to have almost doubled in the past 45 years in Ontario. Bacteria move in water from manure at the soil surface, through cracks and macropores in the soil, into groundwater. Well water in Canada is more likely to exceed drinking water guidelines for bacteria than for nitrate or pesticides.

In McRae and Smith's view (2000), the key challenges facing Ontario agriculture include the need to improve the quality of nutrient management where crop and livestock intensity has increased. They believe that there is a risk that increasing levels of nitrogen, phosphorus, pesticides, and bacteria will be found in agricultural water as the intensity of production continues to grow, especially in southern Ontario.

In order to reduce the use of pesticides, their application should be restricted locally so that areas of the field are left untreated and beneficial insect species such as Syrphidae and Carabidae can recolonize those areas. Wherever possible, selective instead of broad-spectrum pesticides should be employed. Pesticide-resistant pests can be avoided to some degree by careful use of pesticides in a timely fashion, with residual pesticide use where needed (Häni 2001). For instance, from 1981 to 1996 there has been a substantial increase in the farmland areas in Canada that have been treated with herbicides (up by 53%) as well as insecticides and fungicides (up by 78%). Koroluk et al. (1995, 52) indicate that

> although new pesticide products generally pose fewer environmental risks, concerns remain about the impact of pesticides on non-target species and water quality. New biotechnologies, such as pest-resistant crops, and techniques, such as IPM offer opportunities to manage environmental risks associated with pesticide use.

Soil conservation has improved but must continue to do so. In this region, the water resources that agriculture has at its disposal are faced with increasing demand from industrial competitors and agricultural industries themselves, and this will no doubt lead to growing conflicts. Environmental costs are also increasing with respect to greenhouse gases, residual nitrogen, and water contamination by nitrogen, phosphorus, pesticides, and other contaminants.

Greenhouse Gases, Climate Change, and Sustainable Agriculture

Greenhouse gases have been a continuing problem for Canada and are the most serious environmental threat of the twenty-first century. Wherever there has been an increase in cattle, pigs, and mineral fertilizers, greenhouse gases have tended to increase in Canada over the past two decades (McRae and Smith 2000). These gases include carbon dioxide, methane, and nitrous oxide, among others, although the total effect of all carbon dioxide is probably the most damaging. They all contribute to an enhanced greenhouse effect in turn, warming the atmosphere. Environment Canada's senior climatologist, D. Phillips, estimates that the average temperature has risen by 0.6°C over the past 100 years and will probably rise by at least 2°C more over the next 50 years (Phillips 2001). Smit has summarized the major changes as more frequent summer droughts, heat waves, and hot days; less frequent cold waves and fewer frost days; more frequent intense precipitation, hail, and storms; and generally extreme weather conditions (Smit 2001).

In October 2002, the Climate Change Impacts and Adaptation Directorate (CCIAD) of Natural Resources Canada published an assessment entitled *Climate Change Impacts and Adaptation.* It indicates that most agricultural regions will experience warming with "longer frost-free seasons and increased evapotranspiration" (2002, 3), but of course local rates of precipitation, land use, and soil type will modify the effects. Opportunities for some types of speciality crops in southern Ontario, such as apples, will increase. There will be varying effects of greater atmospheric carbon dioxide, benefiting some plants such as legumes but reducing the nutritional values of other crops. Livestock will benefit from warmer winter weather and lower feed requirements, but summer heat may increase the threat to poultry and adversely affect milk production. Soil quality may be negatively affected: "For example, changes in atmospheric CO_2 concentrations, shifts in vegetation and changes in drying/rewetting cycles would all affect soil carbon, and therefore soil quality and productivity" (CCIAD 2002, 7). The greater extremes that will accompany climate change will increase wind and water erosion and could lead to greater flooding and drought. Pests, pathogens, and viruses may migrate north, increasing the range and extent of disease and infestation, for instance, of grasshoppers in Saskatchewan.

Where crop yields are favourable, such as is expected for soybeans, winter wheat, and potatoes, economic benefits should follow. Where greater pest infestations, drought, and flooding occur, economic costs will also go up. Irrigation systems will have to be improved in drought-stricken areas, but biotechnology may help with the creation of more resilient plants and soils (CCIAD 2002).

The international response to climate change has come in the form of the Kyoto Protocol. It would require that emissions of carbon dioxide and other global warming gases be cut to a level 6% below 1990 levels by the year

2010. On 23 July 2001 in Bonn, Germany, 178 nations, without the support of the United States, came to an agreement to accept the treaty rules of the Kyoto Protocol cutting greenhouse gas emissions.

Countries would be allowed to take credit for capturing emissions by various agricultural practices and forests. The agreement sets a price for emissions and determines how much environmental protection can be purchased. It is assumed that the cost of a one-ton carbon certificate will be US$10. At the outset, assuming ratification, gasoline is expected to increase by only 1 cent a litre. As governments promote measures such as improved energy efficiency and renewable energies, local air quality would start to improve, but the targets would also soon have to be ratcheted up to make real improvements (Priddle 2001).

While the federal government announced that its greenhouse gas reduction plan includes the tripling of its current renewable and primarily corn-based Canadian ethanol production (Lawton 2000), most non corn-producing farmers as well as oil companies have not shared the enthusiasm of Canada's National Climate Change Transportation Table (1999) for blending ethanol with gasoline or developing renewable biodiesel fuels, despite the fact that ethanol-blended fuels are the most cost-effective way of reducing greenhouse gases (Klupfel 2000). Klupfel has outlined the considerable advantages that would accrue to farmers and to the mitigation of greenhouse gases if Canada were to move from an economy based on fossil fuels to one based on crops: "The crop-based concept essentially builds on replacing fossil fuel–based feedstocks (including coal, oil, and natural gas) with feedstocks derived from crop matter (such as ligno-power, commodity chemicals and fuels, specialty chemicals, and other materials)" (2001, 12).

Conservation tillage has increased and contributed to some reduction in carbon dioxide emissions, but emissions in general increased by 4% from 1981 to 1996. Although a specific target for agriculture has not been met, further intensification of agriculture will make it difficult to reach the national Canadian goal in the 2008-2012 period of reducing emissions to 6% less than 1990 levels (Desjardin and Riznek 2000).

The Canadian government's objective is to encourage the agricultural sector to "reduce greenhouse gas emissions and minimize the economic impacts associated with climate change adaptation."[4] The agricultural sector can reduce greenhouse gas emissions and promote carbon sequestration through such things as reduced dependency on fossil fuels, use of renewable fuels, increased soil conservation, and other best management practices. In the meantime, farmers will have to adapt to climate change as best they can.

---

4 University of Guelph Farming Systems Research, online at <http://www.uoguelph.ca/OAC/FSR> (retrieved 15 May 2001).

Canada signed its intention to ratify the Kyoto Protocol in October 2001 and, despite opposition from the governments of Alberta and Ontario, ratified it in 2002.

Water Quality

Within rural areas, agriculture uses the most water. It also has been associated with declining water quality. Agriculture affects water quality through "the movement of sediments, nutrients, pesticides, and pathogens off farmland and into water by surface runoff, leaching into groundwater or tile drains, or release to the atmosphere" (Harker et al. 2000, 27).

Agriculture can also increase the risk of water contamination through increased bacteria associated with livestock production as well as increased nitrogen and phosphorus from fertilizer. Soil testing provides a clear indication of the nature of fertilization required. The presence of calcium, for instance, influences soil pH, and is the basis for diseases, pests, and plant growth. A lack of potassium and phosphorus can promote some pests and diseases. Excessive nitrogen fertilizer promotes some plant diseases and pests while reducing the plants' resistance. "Organic fertilizers may increase insect species diversity and densities whereas composts inhibit some diseases of seedlings. Abstaining from both pesticides and N fertilizers in the first 3-5 metres of the field may enhance species richness of weeds and the associated beneficials" (Häni 2001, 5). In Ontario, over the period from 1981 to 1996, MacDonald observed an increase in the nitrogen content of water "by at least 1 mg/L on 68% of Ontario's farmland" (2000a, 12), while residual nitrogen increased at least 5 kg/ha on the soil (MacDonald 2000b).

A study conducted for AAFC between 1991 and 1992 with the help of the Ontario Ministries of Health; Environment; and Agriculture, Food and Rural Affairs, as well as the Ontario Soil and Crop Improvement Association, looked at farm drinking well water quality (Goss et al. 2000; Rudolph et al. 1998). In cases where more than half the land area was for agriculture, 4 wells within each township were sampled. Elsewhere, 1 well per township was sampled, leading to a total sample of 1,300 of the 500,000 wells in Ontario. Participating farm households were questioned about how their wells had been constructed and how far they were from pollution sources. No relationship was found between nitrate contamination and the distance from specific point sources (Goss et al. 2000). For bacterial point sources, a significant decrease in wells with coliform contamination was associated with greater distance between the well and feedlot. The occurrence of contaminated groundwater was related to the type, depth, and age of the water well. Contamination occurred more frequently "in dug and bored wells or shallow sandpoints than in drilled wells, regardless of depth, at lesser depths in all wells [and] in older wells, especially shallower, non-drilled wells" (Goss et al. 2000, 64). About 40% of all the wells tested contained one or more of the target contaminants at concentrations above the provincial drinking water objectives. Approximately 34% were contaminated with coliform bacteria (22% had fecal coliform bacteria) exceeding the maximum acceptable concentration. About 14% of the wells tested contained nitrate beyond an acceptable concentration, and 7% had both nitrate and bacterial contamination. Moreover, while up to 11.5% of the wells had detectable levels of pesticides, only 6 wells (0.5%) showed pesticide concentrations in excess of maximum acceptable concentrations (Goss et al. 2000).

Pollution of surface water occurs as the result of agriculture, not only because of the use of animal and inorganic fertilizers but also because of pesticides. Livestock waste has been recognized as a means of increasing soil fertility and improving soil physical properties. However, Samson and colleagues (1992) argue that over the last twenty years, the use of manure in intensive farming systems as well as much crop production has caused major pollution problems.

And as an example of growing agricultural intensification, modern horticulture, including the greenhouse industry, which produces such things as ornamental flowers, fruits, and vegetables, can also cause high nitrate/nitrogen and phosphorus contents, which in turn can destroy fish habitats and affect drinking water quality. As a consequence, managing fertilizer and water has been of increasing concern to Ontario's expanding greenhouse industry. Environment Canada, AAFC, and the Ontario Ministry of Agriculture, Food and Rural Affairs (OMAFRA) are working in concert with the greenhouse industry to reduce effluents such as nitrates, phosphates, pesticides, and microorganisms that are part of the greenhouse production. To assuage the worst environmental consequences of the greenhouse industry, the University of Guelph's Mike Dixon has developed cultural management strategies for greenhouse production so that production quality can be guaranteed in an economical way without negatively affecting the environment. Recycling nutrient systems can now have increasingly sophisticated sensor technology to monitor nutrient solution quality and help reduce pollution.

The increase of nutrient status in natural water that causes accelerated growth of algae or water plants, depletion of dissolved oxygen, increased turbidity, and general degradation of water quality is called eutrophication (Davies et al. 1993; Pierzynski et al. 1994). The algae, which are tiny plants, sink to the bottom when they die, and oxygen in the water is used up as they decompose. The blue-green algae are common in algal blooms, and when they decompose they emit toxins that can kill fish and other animals. In unpolluted waters, growth of aquatic plants, including algae, is limited by the low level of phosphorus. When phosphorus is added to water, more plants are able to grow, reducing the amount of oxygen and thereby causing the death of fish and other aquatic animals (OMAFRA 1994a). Because nitrate is lost in drainage from farmlands, fertilizers have been considered responsible for the eutrophication of natural water (Cooke 1982).

This is important because the nitrate (NO_3) content of water used for drinking affects the health of babies (Davies et al. 1993; Pierzynski et al. 1994). Moreover, if the level rises above about 20 parts per million (ppm) of nitrate/nitrogen, there is a risk of babies suffering from a disorder known as blue baby syndrome. Standard-acceptable concentration of nitrate/nitrogen levels in consumed water is 10 mg/L for humans, and 100 mg/L for livestock (OMAFRA 1994a). Joy and colleagues recently found for Ontario that even though "there was no relationship between livestock densities and water quality ... nitrate and bacteria levels in rural Ontario streams are not improving, and may be affected by growing human populations" (2000, 53).

Phosphorus pollution comes from many sources in urban areas, mainly sewage treatment plants, storm sewers, and industrial sources. In rural areas, phosphorus pollution originates from sewage treatments from small towns, improper septic systems, storm sewers, manure runoff, milkhouse wastes, and eroded soil (OMAFRA 1994a). Phosphorus from farmland has three sources: the farmstead, pastures near watercourses, and cropland.

Justice D. O'Connor, in Part 2 of his *Report of the Walkerton Inquiry* (2002, 10), has recommended

> that there be minimum regulatory requirements for agricultural activities that create impacts on drinking water sources. The objective of these recommendations is to ensure that the cumulative effect of discharges from farms in a given watershed remains within acceptable limits. For smaller farms in areas that are not considered sensitive, I recommend continuing and improving the current voluntary programs for environmental protection.

As a direct consequence of the Walkerton tragedy and as a partial response to Justice O'Connor's recommendations, Ontario is now belatedly attempting to solve problems of excess nutrients through its passage of the Nutrient Management Act (June 2002). Under the new regulations, four categories of livestock operations have been identified in addition to nonlivestock operations that will be subject to a separate requirement for setting up nutrient management plans. The following regulations were proposed:

> Category IV livestock operations would have 300 nutrient units or more, meaning more than 150 dairy cows or 1,800 finishing pigs.[5] Category III livestock operations would have 150 to 300 nutrient units, or 75-150 dairy

5 OMAF, "Nutrient management protocol," online at <http://www.gov.on.ca/OMAFRA/english/infores/releases/2003/032103.html> (retrieved 25 March 2004): "The Regulation applies to operations that generate, store or use, or transport prescribed materials, as well

cows or 900-1,800 finishing pigs. Category II livestock operations would have 30-150 nutrient units, or 15-75 dairy cows or 180-900 finishing pigs. Category I livestock operations would have less than 30 nutrient units, or fewer than 15 dairy cows or 180 finishing pigs. All other agricultural operations, including non-livestock, would be required to submit nutrient management plans in 2008.[6]

Soil Quality

The principal sources of soil erosion by water are sheet and rill erosion (70-100%), and streambank erosion (0-30%) (Soil Conservation Society of America 1983). Using the agri-environmental indicators of risks of water erosion, wind erosion, tillage erosion, soil organic carbon, soil compaction, and soil salinization, Brown and McRae provide an assessment of the degree to which risks of soil degradation have been affected.[7]

Over the past three agricultural census periods, management of the soil has improved. This is mainly because there is now less conventional cultivation and less soil being left exposed. Soil tillage is either conventional (ploughing below the surface and uprooting and disturbing existing organic material), relatively low or conservation tillage (leaving existing soil layers intact), or non-tillage.

Throughout Canada and elsewhere, conventional tillage is giving way to low-till or conservation tillage and even non-tillage, in order to promote soil conservation. Nevertheless, Häni observes that reduced tillage can have the effect of reducing some diseases and pests while increasing others. Stabilization or natural regulation can result as "many soil fauna are increased in diversity and densities by reduced tillage" (2001, 4). In many cases, when farmers shift to lower-tillage systems, they must adjust sowing dates and nutrients that are available to their plants.

"Between 1981 and 1996 the average number of bare-soil days in Canada's agricultural regions dropped by 20%, from 98 to 78" (Huffman 2000). In

as commercial fertilizers that are used in crop production, that are captured by the phase-in requirements of the regulation ... the Regulation will be phased in depending on: (1) Whether the 'new' operation will: apply for a building permit for a structure that will be used to house farm animals; generate manure or other prescribed materials; generate more than 5 NU in a year; be on a separate deeded property, and on land which the person who owns or controls the site has not previously carried out an agricultural operation that generated manure; whether the farm unit is 'expanding' and the number of farm animals is expected to generate 300 or more nutrient units in a year. (2) An agricultural operation that is carried out on a 'farm unit' under the Regulation will be phased in by July 1, 2005 if: the existing farm unit is expected to generate 300 nutrient units or more in a year."

6 University of Guelph Farming Systems Research, consultations page, online at <http://www.uoguelph.ca/OAC/FSR> (retrieved 15 May 2001).

7 J. Brown and T. McRae, "Soil," online at <http://www.agr.gc.ca/policy/environment/soil_e.phtml#background> (retrieved 16 July 2004).

Ontario, however, because of the extent of such row crops as corn and soybeans, which have little soil cover, there was less than 10% improvement. About 7% more of Ontario's cropland was at risk of water erosion in 1996 compared with 1981, even though the overall risk of water erosion in Canada declined in the same period (Shelton et al. 2000).

The most wasteful and environmentally damaging manure application method, broadcasting, is still widely used except in the Boreal Plains and Prairie ecozones. Injection, which has been used to reduce nutrient losses, was employed on a mere 22% of Canada's cropland. Only 60% of Canada's farmers conducted soil tests in 1995, and manure storage continues to be largely inadequate. Pesticide sprayers are still calibrated only at the beginning of the crop season by 68% of farmers. Nonchemical control methods for pests were used on 56% of Canadian cropland, while tillage was employed as a pest control method on 27% of the land (Koroluk et al. 2000).

Ontario's soils are young and relatively shallow, and are therefore fragile (OMAFRA 1994b). Also, in spite of the recent rise in conservation tillage, some of the tillage systems continue to contribute to the process of soil erosion. Although progress is being made towards more low-till systems, Ontario still uses mostly conventional tillage (OMAFRA 1994a). Stream sediment loads are highest in intensively farmed regions, where there is a high percentage of row crops, fine-textured soils, and extensive drainage networks (Driver et al. 1982; Soil Conservation Society of America 1983).

McCullum and colleagues (1995) argue that the indicators used for assessing soil quality must be based on system objectives. The sorts of soil indicators that are often used in sustainability analysis include measures that are site-specific, such as pH levels, texture, and organic matter content. The problem with these, according to McCullum and colleagues (1995) is that they are too site-specific. Soil quality "relates to the capability of the soil for production or provision of other services beneficial to humans such as pollution attenuation" (McCullum et al. 1995, 6). Those indicators provide us with a sense of the soil's ability to meet certain management objectives. Thus, soil that is of good quality for wheat production may not be suitable for other crops. Soil health, on the other hand, "includes factors which may be unrelated to the achievement of management objectives" (McCullum et al. 1995, 6). Soil indicators alone, in the view of McCullum and colleagues, do not sufficiently take into account other socio-economic factors. Suites of indicators must therefore be used in addition to soil quality indicators for the approach to be more reliable.

Biodiversity

As Neave and colleagues (2000) indicate, climate change has a major impact on agricultural habitat and, in turn, biodiversity. The United Nations Convention on Biological Diversity defined biodiversity as "the variability among living organisms from all sources, including terrestrial, marine, and other aquatic ecosystems and the ecological complexes of which they are part; this includes diversity within species, between species and of ecosystems."[8]

Neave and colleagues (2000) developed an Availability of Wildlife Habitat on Farmland indicator, which is related to other indicators such as the Risk of Wind Erosion, Risk of Water Erosion, Risk of Tillage Erosion, Soil Organic Carbon, Risk of Soil Compaction, and Risk of Soil Salinization. What they found, using the Availability of Wildlife Habitat on Farmland indicator, was that from 1981 to 1996 "habitat area decreased for 74% of the habitat use units in the Mixedwood Plains" (2000, 145). Clearly, then, the most severe environmental problem facing southwestern Ontario immediately is the loss of habitat. Despite the fact that southwestern Ontario is about the most productive agricultural region of Canada, it continuously loses land to heavy competition with non-agricultural uses. The expansion of the urban population to the point where more than half of Canada's population lives in this region has jeopardized biodiversity. However, tilling the soil, draining water from wetlands, and using fertilizers and pesticides all impact biodiversity negatively also. Much of Canada's Carolinian Forest, which is the home of many species usually more common far to the south of this region, is within southwestern Ontario. Neave and colleagues (2000) have pointed out that at least 90% of the wetlands have been drained and much of the original Carolinian Forest has been cut down. As a consequence, many of the original wildlife species in the area are severely threatened.

Neave and colleagues (2000) observe that the expansion of cropland by conversion of wildlife habitat to farmland was the main cause of the decrease in habitat availability. As agriculture has become more intensive with the Mixedwood Plains (and also in the Pacific Maritime ecozone in British Columbia), native pastures and woodlots have been transformed into cropland. Thus, whereas several areas saw reduced cropland (such as eastern Ontario) or reduced summer fallow and more low tillage (the Boreal Plains, the Prairies, and Atlantic Maritime), thereby improving conditions for biodiversity, the situation was far worse in the Mixedwood Plains area of southwestern Ontario.

On the plus side, the increased use of low- and no-till cultivation within the area has reduced soil erosion from water and wind, increased soil moisture and soil organic matter, and reduced soil disturbance.

Biodiversity is declining quite rapidly in this Mixedwood Plains ecozone, especially within the Golden Horseshoe around Lakes Ontario and Erie, as a

---

8 United Nations Environment Programme, Convention on Biological Diversity, online at <http://www.biodiv.org> (accessed 5 June 2002).

consequence of both farming practices and urban expansion into rural areas. Wildlife habitat on farmland also continues to disappear due to the destruction of wetlands and the conversion of pasture land into cropland. A decline of 74% of habitat was registered from 1981 to 1996 in Canada's Mixedwood Plains (Neave et al. 2000). There is therefore a need for a greater effort to conserve wildlife habitat, especially through the establishment of riparian buffer strips, particularly between crops and rivers or streams.

Riparian buffer strips, or what Häni (1998, 2001) refers to as zones of ecological compensation (ZECs), especially hedges with herbs and native flora, can provide the resources needed for beneficials to develop, which can help keep pests and diseases in check. "ZEC's should be arranged in a chain-like network and cover at least 10% of the farmland. Such a network is considered to be suitable for enhancement of beneficials, but this approach must be supplemented and supported by" other appropriate measures (Häni 2001).

Energy Use

For the whole of Canadian agriculture, energy input grew by 0.7% while energy output grew by 1.2%. This was due to improved productivity for mineral fertilizers, pesticides, and fossil fuels. Changes in technology, government policies, farm management practices, and weather patterns affect energy use over time. Growing energy efficiency over time reflects growing productivity and energy output given a relatively constant output of agricultural products. Less use of summerfallow and strong productivity yield increases have helped keep the energy balance positive for Canada. On the negative side, the growth in the use of pesticides, especially in the Prairies, remains a major concern with MacGregor and colleagues (2000) calling for an increased effort to find alternative ways of managing pests.

Work on the negative effects of pesticides predates but includes Rachel Carson's 1962 book *Silent Spring,* which raised widespread alarm by describing some of the negative consequences for wildlife of chemicals such as pesticides in the environment. The word "pesticides" is generic and includes herbicides, fungicides, and insecticides. Many of these chemicals have the effect of disrupting endocrine (hormone) functions in wildlife. High ratios of female-to-male seagulls, as well as fish found with both male and female genitals, have been attributed to excessive pesticide use (Steingraber 1997).

From the 1981-85 period to the 1992-96 period, energy use grew by 8% while energy output experienced a 13% increase. Much of this is accounted for by the Prairies, where energy output was far in excess of energy input. In the Prairies, "on average, major crops (net of feed use) contributed 86% of [Canada's] energy output" (MacGregor et al. 2000, 174). As a result, overall efficiency in the input-to-output ratio for energy remains reasonably high. Energy output actually declined by 3% outside of the Prairies (MacGregor et al. 2000).

However, in southwestern Ontario and other parts of Canada with livestock production and such intensive cropping systems as horticulture, energy input grew by 3% while energy output fell by 3%. During the period from 1981-85 to 1992-96, there was a drop in energy input in machinery as well as mineral fertilizer use. There was a decline in energy output in beef and milk in Ontario and Quebec, but a steady rise in both pork and chicken production as well as energy output with a general shift from corn to soybean production and energy output. Thus, energy use and efficiency by agriculture in southwestern Ontario has performed much better than wildlife habitat (MacGregor et al. 2000).

Hormones, Antibiotics, Biotechnology, and Genetically Modified Organisms

There is growing pressure in Europe against the importation of Canadian beef grown with hormones and Canadian pork, which has been produced with the aid of antibiotics. The European Union has recently published a report suggesting that hormone use in particular can cause cancer in humans who consume the beef. This has proven to be a major worry for members of the Ontario Federation of Agriculture (OFA) and livestock-producing organizations. Ontario Pork chair Clare Schlegel has responded to the European report by claiming that "we have the safest food system in the world" (Landau 2000). Her view and the view of most livestock producers is that these European concerns are simply unfair trade barriers dressed up as health issues.

Europeans are also increasingly opposed to the importation of genetically modified organisms (GMOs). It is important to be aware of the consequences of such powerful technology as gene manipulation. Unfortunately, however, very little is known about how these new species interact with ecosystems. Genetically modified organisms may also destroy the natural diversity of ecosystems, making them more at risk of being damaged by stresses (Häni 1998).

North American agricultural officials have argued that this is really a cover for trade protectionism. Many farmers would prefer not to use GMOs but feel that they are forced to do so because of competition with other farmers who use them, as well as pressure from the chemical and seed companies promoting them. Most Canadians oppose the cloning of animals and are nervous about the safety of genetically modified food. Apart from cloning of animals, however, Ontarians tend to be very supportive of biotechnology (Cobb 2001).

Mad Cow Disease: Bovine Spongiform Encephalopathy (BSE)

Mad cow disease (bovine spongiform encephalopathy) was first found in British cattle in 1986 and was later determined to be the cause of 129 human deaths from Creutzfeldt-Jakob disease (nvCJD) (Contenta 2003). The disease causes animals to behave in a bizarre fashion, reducing their milk production, making them aggressive or nervous, uncoordinated, and often unable to get up. It was eventually determined that the cause of the disease in cattle was feeding on ruminant protein and rendered material from other cattle. The sterilization process that rendered feed goes through does not kill the agent that causes BSE. Canada banned the feeding of rendered cattle to other ruminants. In 1993, when the disease was peaking in Britain at close to 1,000 new cases each week, a Red Deer, Alberta, cow was diagnosed with BSE. It had been imported from Britain. Mexico temporarily banned Canadian beef (*Toronto Star* 2003). In May 2003, tests confirmed that another cow had tested positive for BSE. The test was done as part of Canada's active surveillance program. Fortunately, the slaughtered animal had not entered the food chain, but the cost to Canadian beef producers, who had to slaughter over 2,800 more cattle for testing despite not finding another BSE case, has been extremely high as exports to countries such as Japan and the US ceased. At the time of writing, the Americans and Mexicans had reintroduced a restricted form of beef imports.

Walkom (2003) argues that this was a problem waiting to happen because by allowing the slaughterhouses to sell to rendering plants the 40% of the cow that they couldn't use, the beef industry's cheap food goal came at a huge cost. "Britain responded to mad cow disease by banning the use of cow parts in any kind of animal feed. Canada and the US, however, were reluctant to go that far. The renderers objected. So did the farmers. It would be too costly, they argued." They were not allowed to feed ruminant animals rendered feed, but they were allowed to sell it for chicken and pig feed (Walkom 2003). Although poultry and pigs do not develop the disease, a study published in 1998 in the journal *Nature* argued that poultry and pigs may be carriers of BSE (Laidlaw 2003).

Summary and Conclusion

Globalization and changes in technology, population growth, and market demand have often intensified agricultural production, sometimes with negative environmental consequences. Meanwhile, AAFC researchers argue that the potential environmental risks will probably continue to increase as intensification continues, requiring management responses from industry, governments, and consumers to minimize adverse environmental effects (McRae and Smith 2000, 19).

After assessing the value of using the Driving Force and State to assess sustainability at the micro and macro levels of analysis (refer to Häni 2001), this chapter summarized the research dealing with southwestern Ontario that was conducted by AAFC researchers using agri-environmental indicators.[9]

9 Their research, of course, covers Canadian agriculture and the environment as a whole.

These indicators uncovered some improving conditions, such as greater soil cover due to more conservation tillage on the one hand, and increasing bacterial contamination of water and declining biodiversity on the other.

Climate change is now increasingly occurring throughout the world as increasing numbers of extreme climatic conditions combined with warmer temperatures are affecting us due to increase in levels of greenhouse gases such as carbon dioxide, nitrous oxide, and methane. While there are some things that farmers can do to help mitigate the production of these gases, they must learn to adapt to a variety of impacts on their livestock and plants.

Canada has ratified the Kyoto Protocol and all that it entails, so there is now a federal commitment to both mitigating the production of greenhouse gases through, for example, using alternative energy sources and helping farmers adapt to climate change by introducing new crops. Whether this commitment will waver under Prime Minister Paul Martin remains to be seen. The Kyoto Protocol requires Russian ratification in order to proceed, and it now appears that the Russians are holding out for a better offer before finally committing to it.

Although greenhouse gas emissions have increased considerably in the Prairies due to greater livestock production, they have remained fairly stable in Central Canada. While Central Canada has experienced moderate growth in agricultural GDP, the risk posed to its natural capital by soil degradation actually decreased from 1981 to 1996 except for soil compaction, which worsened. Residual nitrogen levels and nitrogen water contamination worsened as well in southwestern Ontario. Agricultural intensification was associated with increasing energy inputs, while outputs decreased, but not substantially, in southwestern Ontario.

A growing concern is the increasing dependency of farmers on a variety of pesticides. While many of the pesticides currently in use are not as toxic as earlier pesticides, the greater use of herbicides, fungicides, and insecticides underscores the need to expand Integrated Pest Management practices.

Crop rotation has long been used as a method of pest and disease control. Especially in the post–Second World War period, continuing economic pressures brought on by overproduction, low prices, and the consequent concentration and centralization, along with the greater availability of pesticides, increased specialization and the disappearance of such ecologically valuable zones as community pastures. While yields increased with greater use of fertilizers and pesticides, beneficial organisms declined and some pest species spread. In many cases, the spread of pests resulted not from the disappearance of beneficial organisms but rather these organisms' spatial separation from the pests that they attack (Häni 2001).

Establishing riparian buffer strips of native flora around crops and between crops and streams, and limiting field sizes in the process, is a way of establishing zones of ecological compensation (ZECs), which favour natural regulation through the promotion of faunistic diversity. Other solutions include planting green manures, overwintering crops, and allowing for better crop rotations to enhance fertility and the organic content of the soil while reducing erosion. These rotations can employ both a "shifting" of a crop to an adjacent plot the next year or "jumping" so that there are no common borders with crops of the previous year (Häni 1998, 2001).

Manure management is the form of farm management that needs the most improvement. The conflict between increasing intensification of agriculture and growing urban incursion into rural areas has often been manifested in fights over farmers' manure management. The Ontario government's new Nutrient Management Act identifies four separate categories of livestock operation based on the numbers of livestock units. Each of these categories, including nonlivestock farming operations, will soon be required to develop and implement nutrient management plans. It is understood that these plans will be able to take into account trade-offs between the environmental effects of the operations and the need to maintain economic viability. Farmers are being encouraged to use conservation best management practices (BMPs) in their resource management, and many may benefit from the discussion of livestock manure systems in Chapter 8 and the computerized decision support system for manure management in Chapter 9.

As part of his strategy to protect water quality, Justice O'Connor (2002) has recommended that all large farms and small farms within sensitive areas be required to develop a water protection strategy, while small farms in less sensitive areas continue to be encouraged to protect the environment through the use of voluntary plans such as the Environmental Farm Plan (EFP). Soil loss through wind and water erosion can also be reduced through the adoption of the BMPs recommended in plans like the EFP and the Grand River Watershed's cost-sharing Rural Water Quality Program (RWQP).

Declining biodiversity within southwestern Ontario is a major threat to many endangered species. It is important for farmers to develop riparian buffer strips between their crops and rivers. They must cease draining wetlands and replenish woodlots with new trees wherever feasible.

This chapter also summarizes some other agricultural issues that are impacting the environment: growing dependence on pesticides, the use of hormones in beef cattle, "extra-label" use of antibiotics in pig feed, and the increasing use of genetically modified organisms within both plants and animals. Many of these issues' effects are not immediately known but the dramatic rise in organic farming during the last Agricultural Census period is a sign that alternatives to conventional agricultural practices are starting to have a much bigger impact than before.

BSE, linked to deaths in humans from CJD, is another environmental problem of serious proportions. It may be due to insufficient vigilance with respect to the relationship between intensive cattle farming and the processing of feed from rendering plants that is slated for consumption by pigs and poultry. The latter may be carriers, or it could be that the process of rendering cattle remains may contaminate feed destined for cattle, which should not be fed protein from other cattle. Clearly, a tightening up of Canadian regulation of rendering plants is a likely consequence of this expensive new case of BSE discovered in 2003.

While this chapter looked at how Agriculture and Agri-food Canada has been assessing the sustainability of intensive agriculture on a macro level, a brief review was provided of Häni's Response-Inducing Sustainability Evaluation (RISE) for assessing the sustainability of individual farms. RISE uses twelve indicators to estimate the pressure on the State of the farm caused by Driving Forces in order to identify methods of optimizing farm management environmentally, economically, and socially. His view is that governments should be subsidizing only those best management practices that move farms towards greater dependence on natural forms of regulation.

Reducing the use of pesticides and fertilizers, moving to low- and no-till cultivation, and putting greater emphasis on ways of promoting natural regulation can help to minimize agriculture's negative effects on agroecosystems.

3
Social Implications of Intensive Agriculture

Glen C. Filson

Evolving technological changes and growing globalization pressures have accelerated changes in the predominant forms of agricultural production as farmers jockey to remain competitive. While Chapter 2 looked at some of the environmental consequences of our increasingly intensive form of agricultural production, this chapter surveys some of the main social consequences of this industrialization trend.

After discussing the industrialization and intensification of agriculture that have accompanied the latest globalization phase, this chapter reviews a sampling of recent coverage of agricultural issues in major Canadian and Ontario newspapers. A review of the mainly negative impacts of American industrialized intensive agriculture on rural community well-being follows. The importance of class structure in understanding matters of equity and community well-being is raised next. A summary of some of the main factors impacting the quality of life of nonfarm rural people is then provided so that a more comprehensive picture emerges of farm and nonfarm rural interaction in southern Ontario.

Industrialization, Intensification of Agriculture, Food Regimes, and Globalization

Troughton (1982) identifies the rise of agricultural industrialization with the late 1920s, when an emphasis on lowering the cost of production led to the development of common agro-technologies in state and collective farms in Eastern Europe and more capitalistic farms in Western Europe and North America, thereby helping to integrate agricultural production into the total food-industrial system. Features of these new agricultural systems included growing dependence on manufactured farm inputs and the rising productivity of farm labour. This began to replace the family farm, which had characterized the second, "subsistence to market" period of agricultural development, which had existed since about 650 AD and had eventually elevated individual family farms as the ideal forms of production and, to a lesser extent, consumption. In turn, this second agricultural period of history had followed upon the first, which had begun around 10,000 BC mainly in Europe and Southeast Asia as subsistence agriculture steadily replaced hunting and gathering, enabling population growth in its wake. The creation of an urban consumer market for food was critically important in the second agricultural period.

Developed capitalism's third agricultural revolution experienced several developmental waves, including (1) mechanization involving steam and petroleum-based tractors in the 1890s, (2) chemical farming in the 1950s, and (3) food manufacturing in the 1960s and 1970s, first in North America and then in Western Europe: "Taken together, all three phases have brought about the 'industrialization of agriculture'" (Bowler 1992, 11).

Agricultural industrialization is most obvious in intensive livestock production, especially of poultry, swine, and beef feedlots. As an example of industrialized poultry production, Bowler cites the southwestern Ontario "counties of Huron, Lambton, Wellington and Oxford," especially with their linkages to adjacent feed manufacturers and poultry processors (1992, 13). The situation characterized by rising levels of capitalization via purchased nonfarm inputs combined with rising outputs per hectare is known as "agricultural intensification."

Statistics Canada data from two recent census periods show how livestock intensification has been occurring in Ontario as fewer farms produce substantially more livestock. Thus, total production has increased as the numbers of operators have been cut almost in half within the past thirty-five years. There are also fewer small producers and more large-scale operators. While the numbers of Ontario pigs and chickens increased in the five-year periods by 22% and 22.5%, respectively, even greater increases took place in the numbers of pigs and chickens per farm (see Tables 3.1 and 3.2). Despite a small decline in the total number of cattle, their numbers per farm increased as well.

Two associated processes are important: *concentration* and *specialization*. Concentration occurs with the increasing linkage of farming with food processing as larger and larger farms develop guaranteed markets for their produce and processors obtain assured supplies of produce. Specialization occurs as more of the total farm or regional output is accounted for by particular products (Bowler 1992).

Another approach to periodizing history has been developed by Friedmann and McMichael (1989), Le Heron (1993), and most recently Friedman (2002). Friedmann and McMichael (1989) argue that the first truly international food regime had been established during the colonial period of mercantilism, when the production of the colonies was integrated into domestic European economies, which needed food for the growing working classes that were emerging from industrialization. The second food regime was created

Table 3.1

Numbers indicative of growing intensification of Ontario farming

Farm characteristics	1986	1991	1996	2001
Average hectares[*]	78	79	83.1	91.5
Average pigs per farm[†]	241	310	418	695
Average cattle and calves per farm[†]	62	64	68.5	75.9
Laying hens, 19 weeks and older[‡]	828	1040	1,340	1,603
Number of farms[**]	72,713	68,633	67,520	59,728[†]

Sources:

* Statistics Canada, "Total area of farms, land tenure and land in crops, provinces," online at <http://www.statcan.ca/english/Pgdb/econ124g.htm> (retrieved 26 March 2004).

† Statistics Canada, "Livestock, provinces, 1997, 2002," online at <http://www.statcan.ca/english/Pgdb/econ131g.htm> (retrieved 26 March 2004).

‡ Statistics Canada, "Poultry inventory, provinces," online at <http://www.statcan.ca/english/Pgdb/econ109d.htm> (retrieved 26 March 2004).

** Statistics Canada, "Poultry, provinces," online at <http://www.statcan.ca/english/Pgdb/econ132g.htm> (retrieved 26 March 2004).

Table 3.2

Number of animals on livestock farms in Ontario

Type	1966	1996	2001
Number of census farms	109,887	67,520	59,728
Cattle (all ages)	3,136,956	2,228,579	2,140,731
Milk cows	908,699	404,797	363,544
Pigs (all ages)	1,935,595	2,832,082	3,457,346
Chickens (all ages)	25,308,639	35,596,946	43,624,696

Sources: for 1966 and 1996, Keddie and Wandel 2002; for 2001, Statistics Canada, Livestock, provinces, online at <http://www.statcan.ca/english/Pgdb/econ131g.htm> (retrieved 1 April 2004).

following the Second World War and was dominated by American agri-businesses at a time when the main concern was to stabilize world markets in the face of possible market collapse and anti-capitalist revolutions. The US had already established integrated agro-industrial complexes so it was feasible to extend those networks throughout the world that it came to dominate after the Second World War. Despite continued subsidization and consolidating corporate control of food chains favouring larger and more industrialized farms, growing trade liberalization has resulted in still lower food prices during the latest food regime (Friedmann and McMichael 1989; Friedmann 2002).

Rural communities here and abroad are being increasingly influenced by the process of globalization, which has placed considerable pressure on them to change their economic, social, and environmental components in response to increasing population, improvements in information technologies, and changing trade relationships (Wall 2002). The pace of globalization increases as production becomes more internationalized, capital becomes more mobile, nation-states are eroded, Internet communication technologies become widespread, and ideology increasingly emphasizes trade and investment liberalization, privatization, and deregulation (Laxer 1995). Thus, a part of globalization involves the rise of "a 'corporatist' framework [which] has also been proposed as a description of the way in which the state, farm groups and agro-industrial capital appear to have worked in coalition to support the development of agricultural industrialization" (Bowler 1992, 27).

Le Heron and van der Knapp (1995) have also contrasted the earlier long capitalist boom from the 1940s to the 1970s, where the "state's interventionary practice favoured *national* capital and labour over either regional or global capital and labour," with the much heavier involvement of transnational corporations in the more recent period as integrative forces within the global economy. Within the latest food regime, there has been growing integration of certain segments, including vegetables, durable foods, and fresh fruit.

Two models of agriculture are now emerging as a result of these global phenomena. The dominant form in developed capitalist countries is the industrial model, which is based on high input/output, intensive farming that focuses on food quantity, penetration by international capital, and elaboration of biotechnology. The alternative model, which is gaining adherents, emphasizes sustainable agriculture and quality food production (Bowler 1992).

Returning to the local Ontario situation, it is interesting to see how the rise of industrialized intensive agriculture is perceived by the press.

Local Social Issues Arising from Intensive Farming Practices

Just after the *Escherichia coli* O157:H7 outbreak in Walkerton, Gomes (2000) conducted a content analysis of agricultural articles in the *Ottawa Citizen,* the *Toronto Star,* and the *National Post/Financial Post*[1] over the previous five years dealing with recent agricultural changes and the rise of intensive swine production in particular. She began by identifying a series of adjectives, adverbs, and nouns that depict agriculture in a negative or positive light. She then looked for articles on animal rights and animal welfare, large-scale operations, antibiotic use, farm odour, groundwater quality, and manure management. Each issue was further broken down into positive or negative words/arguments.

Despite the fact that the bacterial source cattle farm near Walkerton was only medium-sized, the Walkerton tragedy clearly brought the large-scale

1 She also reviewed the *Globe and Mail, Maclean's* magazine, and a few other prominent local newspapers less intensively.

farming issue to the public's attention and made it part of a public agenda (Gomes 2000). Typical of proliferating stories in this vein was one entitled "Concern grows about pollution from mega-farms: fight against huge factory operations gains attention as manure suspect in outbreak" (Kilpatrick 2000).

The Walkerton tragedy brought beef feedlots into the national spotlight, and the hog industry has been tarred with the same brush. For example, "factory farming has also radicalized the country's multi-billion dollar hog industry in Ontario, Quebec, and the West ... just two per cent of Ontario's hog factories accounted for nearly a quarter of the 5.6 million hogs produced" (Nikiforik 2000). Thus "factory farming" has been vilified in many newspapers and newsmagazines in Canada since the Walkerton tragedy (Gomes 2000).

One of the consequences of the public outcry about Walkerton was the passage of the Nutrient Management Act (NMA) by the Ontario Legislature in June 2002. This act initially required all livestock and crop farmers, large and small, to develop nutrient management plans by 2008. This deadline has now been dropped for small operations after considerable lobbying from farm organizations saying that the costs of complying with the legislation would be particularly onerous (Avery 2003).

Many recent newspaper articles have referred to large-scale operations as being corporate with little mention of the fact that most farms are still family-owned and operated[2] even if the majority of production is from commercial farms. One of the few positive arguments made in the major newspapers in favour of these large-scale farms was job creation. The example of a 1,800-sow barn being built near Ottawa should, according the township's reeve, "benefit local businesses and create 11 jobs ... boosting business at local feed supply stores" (Egan 1997). In many cases, the *National Post/Financial Post* had the most positive tone when dealing with large farms. This was exemplified by an article written about Big Sky Farms Inc. of Humboldt, Saskatchewan, which was planning to build "three large-scale facilities, each with a capacity to turn out 58,000 market hogs annually," all to improve the local economy, create economies of scale, and meet consumers' growing demand (Shecter 1998).

Gomes, however, also found many articles that took aim at large-scale livestock (for instance, Egan 1997) and even cropping operations. These stories complained of the "destruction of rural scenery, possible environmental damage, bias of large-scale farmers, lax regulations, concern over problems in other geographical locations, the perceived corporate nature of hog production, and intensified concerns following the recent *E. coli* outbreak in

---

2 Ontario Farm Animal Council, "Frequently asked questions," online at <http://www.ofac.org/faqs.html> (retrieved 7 June 2000).

Walkerton, Ontario" (Gomes 2000, 3). There have also been major criticisms of antibiotic use in pork production (such as Flack 1998).[3] While some articles defended the industry, most have seen antibiotics as problematic and have assumed that their use is excessive (Gomes 2000).

Even though the expansion of large-confinement swine operations in Ontario has not occurred at as fast a rate as in North Carolina and parts of western Canada, this distinction has escaped most newspapers. According to Lindgren (1999) of the *Ottawa Citizen*, for example, demands for strict legislation have arisen since "lax regulations on large-scale pig farms led to serious water and air pollution problems in the Netherlands, North Carolina and Quebec."

With respect to animal welfare and animal rights,[4] Gomes indicated that "no articles with a positive overall tone were found in any of the newspapers. Furthermore, articles that dealt with this topic seemed to focus on the life of pigs in so called factory farms" (Gomes 2000, 3). A typical title was "Intensive pig life is a tortuous one" (Khimji 1998). Gomes's quantification of her content analysis revealed a more negative than positive overall portrayal of agriculture by these prominent newspapers.

Some of the most serious concerns of nonfarm rural people and conflicts between them and the farming community have developed because of the growth of large-confinement swine production. Complaints in North America about pork production, particularly regarding large-scale facilities, usually emanate from (1) new rural residents, (2) longtime rural residents, and (3) neighbouring farmers (Abdallah and Kelsey 1996). With respect to some issues, neighbouring smaller-scale pig farmers are often very critical of expanding operators because of a perceived or real threat of the expansion to their livelihood activities (Filson and Friendship 2000b).

Swine operations often encounter more human reaction than other livestock operations (Ritter 1989; Spoelstra 1980). Complaints about water pollution, odour, and manure disposal are evidence of increased conflict between farmers and nonfarm residents in the surrounding areas. Rural residents have demanded everything from a complete restriction on operations to pressure for more control to minimize the negative impacts on the environment and personal welfare (Arcand et al. 1999). Reaction to the expansion of new swine barns in Ontario has varied from expressed concerns about odour, water quality, and how it is going to change their lifestyle and their

3 Extra-label use, wherein hormones and antibiotics are routinely used with livestock in ways not recommended by manufacturers, is now being investigated by Health Canada because of the danger of chemical residues in food and their general release into the environment (Freeze 2002).

4 None of this is meant to deny that there are important animal welfare issues and problems; however, the FSR group's work did not directly address these questions.

neighbourhood to the formation of organized public groups in several counties to prevent or to restrict the swine facilities from being built. At times this has forced hog producers to modify their plans, causing unexpected delays and extra costs (Toombs 1997).

In order to understand what was happening in two hog-producing counties of southwestern Ontario, Filson and Friendship (2000a, 2000b) undertook a detailed study of the views of producers and nonfarm people alike about pork production in Waterloo Region and Perth County in two recurrent surveys. The first was conducted with 537 hog-farm and nonfarm rural people in 1999, and the second was conducted with an additional 465 farm and nonfarm rural people in 2000. Nonfarmers living close to pig-producing facilities were most concerned about (1) possible contamination of groundwater due to inappropriate pig manure application, (2) the implications of the increased size of hog operations, (3) odour from pig production, (4) storage of pig manure, and (5) expansion of large-confinement swine production.

The Tendency for Quality of Life to Decline as Agriculture Industrializes

Wall (2002) has observed that the effects of globalization and the economic changes that are taking place in rural Ontario are likely to impact quality of life negatively. Values are less cohesive now that social values are less dominated by the agri-food sector, rural populations are aging, voluntary organizations face declining memberships, and land use conflicts are increasing.

Defining social capital as the "norms and networks of social relations that build trust and mutual reciprocity among community residents, social organizations, and civic institutions" after Potapchuk and colleagues (1998, 5), Owen and colleagues (2000) argue that social capital in many rural Canadian communities has been significantly diminished. As rural residents have resorted more and more to conducting their business in larger centres, Owen and colleagues (2000) argue that their rural community social bonds have been weakening. Fuller (1994) also claims that the downside of the concentration of agricultural production and diminished number of farmers is increased social distance.

Except for dairy farmers, who usually have relatively less recourse to off-farm work, close to half of Ontario's farm households have someone working off the farm (Filson 1993). The amount of off-farm work is highest for beef producers (Keddie and Wandel 2001). With a sample of western Ontario farmers, McCoy and Filson (1996) learned that an even lower proportion of farmers with off-farm work are satisfied than those who work only on the farm. The highest levels of dissatisfaction were identified among farm women who were employed off their farms. With their "triple day," these women rarely have much time for leisure and their perceived quality of life suffers.

Also, Smithers and Joseph (2000) note that off-farm work and debt are positively associated.

On the positive side, improved production techniques have now made it possible for many family farmers to increase their scale of production without having to resort to hired labour. This has led to a higher perceived quality of life for some farm operators without the attendant alienation that agricultural workers often experience (Filson and McCoy 1993). As Giles and Dalecki have observed, "the presence of large numbers of hired farm workers is more likely to be negatively related to socioeconomic well-being than is the presence of large-scale farms" (1988, 52).

The most significant rural sociological research thrust over the past two decades has centred around empirical assessments of the so-called Goldschmidt Hypothesis. Based on research from 1944 to 1978, Walter Goldschmidt studied the growth of agribusiness in the central California valley by comparing Arvin, a community with relatively large capitalist farms, with Dinuba, a community with smaller-sized "family farms." He found that farmers' quality of life – as measured by the quality of the schools, the degree of poverty, median family income, and the volume of retail trade – was inversely related to the scale of operation in these California communities (1978a, 1978b). His hypothesis was, therefore, that industrialized farming is detrimental to rural communities. A major problem with Goldschmidt's approach, however, is that it was not longitudinal but instead based on a static, cross-sectional analysis (Goss 1979). Saha also points out that, like so much social scientific research, "farm structural variables have been taken as independent variables" (2003, 15), as rural community impacts are assumed to be the result of changing farm structure, not the consequence of rural community changes.

Lobao (2002) has recently summarized thirty-seven studies that have tested this hypothesis since the 1970s, when the first, quantitative phase of testing of Goldschmidt's hypothesis began. She has observed that whereas the 1980s were primarily absorbed with identifying conceptual and empirical limitations of the 1970s studies, the most recent decade has seen a renewed concern about industrialized farming. Once again there has been a proliferation of quantitative and qualitative studies of the effects of large-scale farming on rural community well-being. Lobao's review of these studies determined that in thirty of the thirty-seven studies of the impact of the growth of industrialized farming, at least some detrimental effects on community well-being were found. While some positive effects, such as lower food costs, greater regional income, and improved property values, have been discovered, many more detrimental effects have been found, such as less total community income, greater income inequality, farms less involved in communities, more "lower-class" farm personnel, lower average farm family income, reduced property values, and negative odour impacts. Lobao concluded that "increased vertical integration of corporations into farming through either direct production or contracting arrangements, is altering rural livelihoods and exacerbating spatially uneven development" (2002, 1).

Declining rural community well-being due to agricultural industrialization can eventually be offset by the fact that the rural community's socioeconomic well-being will increasingly be dependent on nonfarm activities such as exurbanite rural settlement, improved transportation, and the diversification of manufacturing (Saha 2003). While most of the sociological literature attributes a declining rural community sense of well-being to the industrialization of agriculture, some do not, such as Swanson (1990). Also, if there is considerable off-farm employment available, this will mitigate the effects of the intensification of farming by providing small farm operators with additional income (Swanson 1990; Goddard et al. 1993; Saha 2003).

With the rising intensity of farming and greater corporate control of farms, a growing number of farmworkers instead of small farm owners is seen. This can have negative consequences for rural social structure, such as declining incomes, educational levels, community participation, and overall rural community viability (Buttel 1983; Filson 1996; Saha 2003), and it can sometimes produce a bimodal distribution of income (see Chapter 9). As agricultural industrialization occurs, the dominant position of agriculture in the development of rural communities fades as other forms of economic activity rise in relative importance, as Gertler (1999) found for many Saskatchewan communities. The effects vary as a function of such things as the religious and cultural composition of the rural community, the relative importance of agriculture to the community, and the extent of both in-migration of exurbanites and out-migration of youth.

Large-scale farming in California has depended heavily on the use of migrant labour. This is increasingly the case with many tender fruit, vegetable, and especially tobacco farms in southwestern Ontario, but it may not apply to some other commodities. Within the fruit, vegetable, and tobacco farms where they are usually employed, Mexican and Caribbean labourers who participate in Canada's Seasonal Workers' Program are not protected in the same ways that indigenous workers are. They cannot leave one employer to work for another or do much to improve working conditions that are among the worst in Canada.

While most Ontario farmers feel that their quality of life compares favourably with that of others in the province, most farmworkers' perceived quality of life has not been studied to the same extent, although farmers have been known to complain about the difficulties of obtaining and keeping Canadian farmworkers. Dairy farmers, who credit their supply managed system with providing them with a relatively good, stable income, have among the best overall farmer-perceived quality of life (Filson and McCoy

1993; Filson et al. 1998). In general, farmers whose quality of life is perceived to be better than that of the average Canadian tend to have relatively more education, earn a sizable net income, and employ "farmhands" to help them with their work (Filson and McCoy 1993). Farmworkers, on the other hand, work long hours in often dangerous circumstances for low wages.

Farmers and nonfarmers alike recognize that the scenic and aesthetic qualities of their environment as well as its physical condition and the air, soil, and water quality affect their quality of life (Caldwell 1994; Filson 1993, 1996; Filson et al. 1999; Filson and Friendship 2000a). This provides farmers with more incentive to introduce environmental best management practices (BMPs). The downside for quality of life, however, arises from the costs of complying with environmental standards in both economic and labour terms (Filson et al. 2001). Unrecompensed costs borne by farmers alone adversely affect their pocketbooks and quality of life despite the fact that their failure to deal with environmental problems could adversely affect them and their neighbours. If these costs are sufficiently high, the farmers may be forced out of business by their efforts to comply with environmental regulations and/or guidelines. There are therefore important trade-offs between the farmers' economic and environmental circumstances regarding their willingness to do the right thing with respect to the environment. Their competitive position and even their survival could hinge upon environmental constraints imposed by their municipal and provincial regulators (Cruise and Lyson 1990; Filson 1993, 1996; McCoy and Filson 1996).

Ontario's Farming Classes

In a gradational sense, clearly there are farmers with relatively more income, education, and scales of operation than others, so that in this sense there are obviously upper-, middle-, and lower-class farmers. Different types of farms have different requirements for hired labour and different needs to engage in off-farm labour, so these classes vary by commodity group.

From a discrete social class point of view, Bowler (1992) argues that capitalist accumulation tends to produce a capitalist landowning class of farmers operating large wage labour farms largely run by farmworkers. Friedmann (1986) points as well to a simple commodity producing (SCP) class of family farmers based largely on family labour with only occasional use of wage labour, although Bowler (1992) feels that the concept of SCP underestimates the extent to which small and medium-sized farmers hire workers while many large-scale and capital-intensive farms can still be family farms. He also points to the vigorous debate over the evidence for a "disappearing middle" of medium-sized farms in an increasingly polarized farm structure.

Following Filson's discrete classes (1983, 1993), there are two main types of farming systems in Ontario: (1) the majority of family-owned and operated farms, which produce a minority of agricultural production without regularly employed hired workers, and (2) commercial or capitalist farms, which employ farm managers and workers and also account for the vast majority of agricultural production. The latter, more intensive farms include the farm owner(s), managers, and workers, and often some family labour as well. Family farms exist in virtually every farm commodity while commercial farms are more common in some commodities, such as tender fruit and vegetables, cut flowers, pork, and poultry.

Considering that there are about 100,000 farmworkers in Ontario (MacCharles 2001) and fewer than 60,000 farms, of which 24,013 reported paid work, there are now more farmworkers than farm owners in Ontario even though 911,030 of the farmers' paid weeks of work per year out of 1,376,166 were seasonal.[5]

A survey funded by the Ontario Ministry of Agriculture and Food (OMAF) in the early 1990s was conducted across Ontario to assess how discrete social classes of farmers felt about sustainable agriculture (Filson 1993, 1996). The survey found that the more intensive, commercial farmers were most opposed to government intervention to enhance agricultural sustainability, feeling that their operations were already largely sustainable (Filson 1996). These commercial operators also expressed the greatest support for the use of various agricultural chemicals (Filson 1993). Conversely, those most receptive to government regulations and interventions designed to enhance rural equity, stability, productivity, and sustainability tended to be the small operators with off-farm employment (worker-farmers), particularly if their main commodity was beef. Farmers without employees usually fell somewhere in the middle on these questions (Filson 1996). Evidence of significant conflict between large and small farm operators was found within pork production by Filson and Friendship (2000a).

Conflicts over Rural Land Management and Use

The growing intensity of farming operations has also been accompanied by growing conflict between nonfarm residents, now usually in the majority within Ontario rural communities, and the reduced numbers of farmers. Owen and colleagues (2000) identify seven causes of increasing rural conflicts: more clustered, larger farms, changes in rural demographics, changing societal expectations, increasingly organized representation of interest groups, changing governmental role, diminished social capital, and continuing globalization. Smithers and Joseph (2000) believe that the increasing conflict is arising from the delinking of farms from the rural communities within which they reside.

5 Statistics Canada, "Paid agricultural work, provinces," online at <http://www.statcan.ca/english/Pgdb/econ115g.htm> (retrieved 26 March 2004).

Troughton describes the delinking that has taken place: "With the reduction of agriculture, these functions and linkages have declined and the rural hamlets, villages and small towns have become functionally and socially separated from the extensive landscapes they once served" (1998, 4). Rural communities have been profoundly affected by the reduction in the number of farms and the replacement of local rural-agricultural supply with agribusiness contracts so that traditional functions and linkages that used to exist between farms and rural communities have largely faded away. This has weakened the distinctiveness and countryside characteristics that had prevailed for over a century. Exurban amenity and residential functions have allowed some rural communities to survive (Troughton 1998), but the decoupling of agriculture from rural Ontario communities has increased the conflict between exurban residents and the remaining farmers.

In response to these changes, Owen and colleagues (2000, 482) advocate the use of interactive conflict resolution (ICR) as a way of increasing "community social capital by creating platforms for planning and action, and by creating an increased stock of individual skills and knowledge." Smithers and Joseph (2000, 265) point instead to several ways of relinking rural communities and the often larger farms in their midst so as "to foster communication, collaboration and, ultimately, mutual sustainability." Most recently, however, Joseph and colleagues, in a study of rural New Zealand, which built upon the Smithers and Joseph work in Ontario, have concluded that "the distinction between de-coupling and relinking, while conceptually attractive, is empirically problematic" (2001, 25). This is because of the complex and ambiguous mixture of de-coupling and relinking that occur between agriculture and their particular rural communities.

Conflicts continue to exist, however, and not only between farmers and nonfarm rural people; as can be seen above, there are conflicts and divisions within the farming community itself. Older farmers, for example, often favour severances of part of their land for sale to incoming urbanites so they can afford to retire on their land, whereas this practice is condemned by other farmers who want to preserve scarce agricultural land. This problem is particularly acute in the southwestern part of Ontario around Niagara, where urban sprawl and land severances have reached significant proportions.

Even though Friedmann and McMichael (1989) say that local agricultural practices are dictated by an international "food regime" that is not the result of local choice, they believe that "relocalization" provides part of the antidote to the problems that globalization creates. Friedmann (2002) thinks that these systems will be able to assuage the ecological and social costs attached to the food crisis and redeem the fantasy that rural living has excellent quality-of-life benefits only by relocalizing food systems. To accomplish this relocalization, she argues that food and agriculture need to be reorganized around health issues.

There are also other ways of relinking farmers and nonfarm rural people. Caldwell (2000) believes that much more can be done to promote "good neighbour" policies at the community level between farmers and nonfarmers. And Fuller (2002) asks how we can engage the 80% of the farming population to produce public goods, thereby helping to re-engage many of the farmers who have become increasingly separated from the majority of rural residents. It may now be time to engage the majority of farmers in land stewardship programs that produce public goods that enhance the public's quality of life. Not only does the public have an interest in farmers' adoption of best management practices to protect the environment but the relatively less productive farmers could be paid to manage their land to produce such public goods as agri-tourism and landscape maintenance (Fuller 2002). The European Union has already recognized and funded what they describe as farmers' multifunctionality. They are now shifting from commodity support to agri-environmental payments that reward farmers for land stewardship and the enhancement of agri-tourism (Bryden 2002).

Another way of relinking farms with local communities is through the development of community shared agriculture (CSA). Kneen (1989, 1990, 1999) has championed CSAs and other ways of reducing the physical distance between where food is produced and consumed. In the Niagara region, for example, various forms of alternative agriculture such as community shared agriculture and farmers' markets are helping to create local job opportunities, provide other marketing opportunities for local farmers, and raise the awareness of food consumers by putting them face to face with local farmers and their produce. "The basic concept of community shared agriculture (CSA) is that members of the community can share in the risks and benefits of farming by paying in advance for a share of the crops produced, and often by contributing labour throughout the growing season" (Krug 2000, 277).

Probably the most significant change that has taken place in response to the problems associated with conventional agriculture is the rise of organic and/or biodynamic farming. Saskatchewan now boasts 773 certified organic farms, while Ontario has 405.[6] Although "certified organic farmers" are still fewer than 1% of the total, their growth is significant.

For now, however, these alternatives are merely nibbling away at the dominant circumstance of a globalizing food system, the growing intensity of most agriculture, and increasing urban sprawl into the countryside.

Understanding Farmers' Stewardship Behaviour by Class and Demographics

Continuing pressure on air, land, and water resources is occurring due to

---

6 Statistics Canada, "Certified organic farming, provinces," online at <http://www.statcan.ca/english/Pgdb/econ103g.htm> (retrieved 26 March 2004).

further agricultural intensification and urban-to-rural migration. Studies dating back to the 1970s indicate the negative impact that livestock and poultry wastes are having on the Great Lakes Basin (Bangay 1976).

Some religious beliefs may also affect BMP adoption, although more research is needed to clearly distinguish religion and ethnicity from other demographic factors and the effects of farm type and size within specific regions. Nonetheless, Ryan (1999, 193) discovered that "members of the Christian Farmers Federation of Ontario (CFFO) often scored lowest on the individual statements relating to the environment" compared with Ontario Federation of Agriculture members and farmers who were not members of either group. Despite observing that the CFFO's leadership is committed to environmental stewardship, she noted:

> Stewardship or "earthkeeping" implies a use of creation to benefit all people and creatures. The concept is that if we care for creation it will provide us with all we need to survive. This differs considerably with standard environmental ideology that tends to emphasize preservation rather than use (1999, 194).

Bucknell and colleagues (2003) also found that participation in environmental programs by farmers and rural landowners in the Canagagigue Creek region, an area with a large traditional religious population and known environmental problems, has been significantly less than participation within the Eramosa/Speed region, a region consisting of people whose religious views are also mostly Christian but of a more heterogeneous variety. Even if lower adoption of conservation BMPs turns out to be somehow related to strong religious convictions, it cannot be attributed to religious orientation alone, of course, as factors such as farm size and type and demographic factors such as a relatively low educational level are inevitably important as well.

CFFO strategic policy adviser Elbert van Donkersgoed is pleased that OMAF has backed off the requirement of nutrient management planning (NMP) for all farms for the time being, but he still opposes the NMP requirement for small operations in the long run.[7] He argues that "there is a case to be made for nutrient management strategies and plans for our larger and intensive farms. But for the vast majority of farmers, what's the point of writing down their plans and documenting what they have been doing for decades?" His opposition to the regulatory approach to nutrient management may have as much to do with a defence of small family farming as it has to any religious perspective.

---

7 E. Van Donkersgoed, "First identify the threats," *Corner Post,* 9 June 2003, online at <http://www.christianfarmers.org/sub_news_commentaries/sub2_news_com_corner_post/news_com_corner_post.htm> (retrieved 16 July 2004).

In a study of the factors that affected 427 farmers' adoption of soil and water conservation measures in southwestern Ontario, Serman and Filson (2000) discovered that the number of crops cultivated, the size of the farm, the level of gross sales, and formal education were all correlated with adoption, whereas age and years of experience made no difference. Cost was an important barrier to BMP implementation, especially for small-scale farmers. The type of operation also makes a difference, with beef farmers usually having a higher rate of adoption of environmental practices such as fencing, habitat restoration, and the creation of buffer strips than other livestock producers.

Filson (1996) also found different degrees of willingness to participate in environmental programs among different types of farms, with dairy and swine farms being among the least willing participants. Of course, the expenses associated with environmental programs, including such things as manure storage systems and milkhouse waste prevention, can be a bigger deterrent to livestock farmers than associated expenses for nutrient management on farms based solely on cropping systems.

Even though a previous study by Buttel and colleagues (1981) had found that education was not significantly related to environmental concern, Duff and colleagues (1991) argued that education and financial incentives are often seen by farmers to be more effective than other government policies in promoting soil conservation. Although farmers usually recognize that regulations are sometimes needed, Duff and colleagues observed that farmers often prefer a voluntary approach.

While farmers are aware of the need to protect environmental resources, research in Canagagigue Creek and Eramosa/Speed subwatershed (Bucknell et al. 2003) revealed that farmers generally tend to believe that government intervention in agriculture and the Nutrient Management Act (NMA) will hurt them economically. Many would have preferred the continuation of voluntary environmental programs (the Environmental Farm Plan Program and the Rural Water Quality Program), whereas nonfarmers supported nutrient management and other new government environmental regulations without reservations (Filson et al. 2002). Farmers held negative attitudes towards government regulation in general, although they agreed that it is appropriate to have water quality regulations. Farmers in both the Eramosa/Speed and Canagagigue Creek subwatersheds of the Grand River were found to usually have a more instrumental, or use-oriented, view of the environment than nonfarmers, who tended to be more concerned about protecting scenic qualities and biodiversity (Bucknell et al. 2003).

At present, the Farming Systems Research group at the University of Guelph is working with landowners in watersheds where there are known environmental and social problems to try to understand the biophysical nature of those problems, the economic/environmental trade-offs, and the relationships between farmers' motivation to adopt conservation BMPs and their specific demographic and farm characteristics. The analysis of these relationships varies to some degree from one area to another, so FSR analyses like this provide essential information for developing the most beneficial combination of technical and financial incentives to encourage environmental stewardship.

Summary and Conclusion

We are now functioning within a global food system that is increasingly privatized and liberalized, and that demands internationally competitive farmers due to lower food prices and the domination of corporate food chains. Farming operations have become increasingly industrialized, intensified, specialized, and concentrated. This favours bigger, more commercialized farms, especially within branches of agricultural production – such as poultry, greenhouses, and pork production – where capital intensity leads to returns to scale. As the food and environmental crises increasingly come together, there is a need to provide greater public financial support for environmentally friendly agriculture.

To some extent, a polarization has already occurred between large and small producers and between farms and nonfarm rural community members. In turn, this has often been exacerbated by sensationalist news reporting about modern farming practices. Clearly, however, social preferences have evolved to demand a more environmentally neutral agriculture because of agriculture's effects on both the environment and people's quality of life. Most Ontario farmers believe that their farming systems are relatively benign environmentally, although nonfarmers tend to fear the environmental consequences of intensive agriculture, especially with respect to agriculture's impact on the quality of groundwater and their quality of life.

The urban and nonfarm rural public's negative perceptions appear to have taken a turn for the worse, especially since the Walkerton tragedy, and now stand as a major symptom of unresolved tensions in rural southern Ontario. Many nonfarmers have moved into rural Ontario in search of parklike conditions to go along with the beautiful, bucolic farmsteads that they wanted to see but not hear or smell. There has been growing conflict between the increasing numbers of exurbanites and the declining numbers of farmers, who now usually operate larger farms than in the past. Among the many causes of conflicts between rural nonfarmers and farmers are the greater concentration, specialization, and centralization of farm operations and the relative ignorance most neophyte nonfarm rural people display regarding modern farm practices.

This chapter has also considered a major preoccupation of North American rural sociology: whether the growing intensity of our agriculture will lead to the fulfillment of the Goldschmidt hypothesis in Ontario; that is, that the more industrialized our farming becomes, the worse the socio-economic conditions and quality of life in rural communities will be, at least for the growing numbers of agricultural workers. Detrimental effects of industrialized farming include increased income inequality, reduced property values, and reduced farm involvement in rural communities.

Commercial farmers tend to be most in favour of conventional practices, including the use of various chemicals, and are among the least supportive of what is termed "sustainable agriculture." Relatively marginalized worker-farmers with significant off-farm income would like government help to introduce more sustainable practices and protect the small family farmers and simple commodity-producing independent family farmers. The large majority falls somewhere between commercial farmers and small farmers in their views about issues of equity and sustainable farming. The rate of growth of more intensive farms has raised some concerns about water, air, soil, and well-being quality within surrounding rural areas, but the lack of understanding of farming practices by exurbanites sprawling into rural communities has also increased tensions with farmers.

This chapter has also reviewed some possible solutions to the growing separation between increasingly intensive agricultural production and the majority of rural residents who are now nonfarmers. This gap may partly explain the rise of interest in relocalizing agriculture, the focus on healthier ecosystems, interest in ways of improving food safety, and a focus on food quality rather than quantity. These objectives are common to community shared agriculture, organic, resource-efficient, and ecological forms of agriculture. For now, at least, these alternatives are still fairly marginal relative to conventional agriculture, so progress must be made towards enhanced sustainability within mainstream farming systems. Other ways of relinking farming systems with rural communities that deserve attention include government-funded support for farmers' multifunctionality and the creation of public goods for agri-tourism, environmental protection, and landscape maintenance.

Organized farm opposition to the Nutrient Management Act, particularly as it applies to small operations, has saved small operators from having to develop nutrient management plans, at least for the foreseeable future. However, as pressure on farmers to implement environmental BMPs continues to mount, local watershed landowners are being scrutinized more closely as a function of their demographic and farm characteristics, to better understand their stewardship motivation and the optimum mix of financial and technical assistance needed to encourage the implementation of these BMPs.

Part 2:
Framework and Linkages

4

FSR Concepts and Methods for Addressing Social and Environmental Problems

John Smithers, Ellen Wall, and Clarence Swanton

There is now broad consensus that workable definitions of sustainable agriculture must include physical, biological, and socio-economic elements (Acton and Gregorich 1995; Benbrook 1991; Schaller 1990; Smit and Smithers 1993).[1] There is less agreement on just how to merge these in practice, however. Sustainable agriculture generally, and farming systems research specifically, needs practical tools to assess and solve problems related to the sustainability of existing and emerging agricultural systems, especially in an integrative manner (Flora 1992).

Despite the knowledge gained in traditional disciplinary research about physical, biological, social, and economic processes in agriculture, there exists only a limited understanding of how these processes are interrelated in different farming systems, or how these systems might affect the environmental and social health of rural communities. The former issue has been the focus of work by agricultural systems analysts (e.g., Dent et al. 1995; Doyle 1990), but not in the explicit context of sustainability. The latter has been attempted by a number of analysts (e.g., Dumanski et al. 1995; Stockle et al. 1994; Smit et al. 1997), but not in a dynamic systems approach. Few would disagree that the multifaceted and largely philosophical nature of sustainable agriculture must be defined and operationalized in ways that are conducive to "systematic and systemic" description and analysis (Bawden 1995).

Among those who are disadvantaged by the lack of progress on moving sustainability from the conceptual to the practical are the very farm operators whose family and community livelihoods hang in uneasy balance. North American farmers find themselves subject to growing criticism regarding the environmental damage from their practices (Beus and Dunlap 1990; Kelly

1 A similar version of this chapter, entitled "An Integrated Framework for Solving Problems in Sustainable Agriculture," appeared in the now defunct *Journal for Farming Systems Research-Extension* 7, no. 2 (2002): 43-58.

1996; Tisdall 1992) despite the fact that in some regions there have been significant reductions in the use of pesticides, chemical fertilizers, and fossil fuels coupled with increased adoption of conservation-oriented soil management practices (Surgeoner 1996).

It is incumbent upon sustainable agriculture researchers and advocates to work towards implementing a clear conceptualization of the term and then contribute to a process for evaluation that might, in some instances, include the development of acceptable thresholds or targets for farm operators to meet. In that way, agricultural analysts, farm operators, and their critics can evaluate farm operations and determine whether or not progress has been made (Johnson 1996).

The purpose of this chapter is to present a problem-solving framework for researching and alleviating problems in sustainable agriculture. The approach and concepts described here are derived from previous scholarship in a variety of research and environmental management paradigms, and from recent deliberation and scholarly debate among members of an interdisciplinary farming systems research (FSR) program at the University of Guelph.

The broad goal of the University of Guelph's Farming Systems Research and Extension (FSR/E) project has been to develop and test an approach for assessing the sustainability of Ontario's farming systems. An initial objective of the project was to develop an analytic, indicator-based framework for assessing the sustainability of current farming systems and for exploring the impacts of changes and/or stresses on those systems. The desired outcome was an improved understanding of the prospects for, and barriers to, sustainability in various farming systems and the delivery of assessment tools that would be of use to researchers, extension personnel, farmer operators, and policymakers in improving Ontario's farming systems.

The framework establishes a sequential, systematic, and strategically iterative approach that parallels the general farming systems research process, where *objectives* are established as an initial step followed by *information gathering* and *evaluation* (Petheram 1986). However, as Flora (1992) has noted, the conventional "steps" of FSR/E need to be modified to reorient research towards sustainability assessment and intervention as opposed to the historically dominant purpose of increasing productivity. The four main components of the tradition – diagnosis, design, on-farm trials, and extension – have been integrated into the FSR framework, emphasizing the logical, sequential nature of the research and the potential for change, both planned and incidental, in the biophysical and socio-economic dimensions of farming systems.

In keeping with Baker's assertion (1993) that farming systems research should be more attuned to agricultural policy issues, the integrated FSR project has been interested in working not only with community-based stakeholders and commodity groups but also with policymakers at various jurisdictional levels. In North America, for instance, the state has maintained a high profile in pursuing sustainability issues, although its effectiveness in doing so has been questioned (Hall 1997). Thus, there are opportunities and needs for policy-relevant contributions from agricultural research of the type described here. There have also been several initiatives designed to promote more acceptable practices where the impetus has come from producers and associated farm groups themselves via partnerships with public institutions.

One of the most significant is a joint government and industry effort called the Ontario Environmental Farm Plan (OFEC 1993), which has been reviewed and adopted by over 20% of the provincial farming community. By the summer of 2000, roughly 18,000 of Ontario's farmers had gone through the farm plan and modified their operations on the basis of twenty-three criteria pertaining to measures for everything from improving soil and water quality to enhancing the natural environment of the farm property (see Chapter 11). Over half of these farmers have taken the further step of submitting their changes to a peer review process, thereby receiving official documentation regarding the environmental aspects of their operations.

Finally, we note that our focus is on developed industrialized agricultural systems. Hence, some of the caveats associated with "traditional farming systems research" are avoided. As noted by Berdequé (1993), farming systems research has most often been concerned with small-scale, resource-poor farmers in marginal areas of developing countries. There, problems with credit availability, weak markets, powerless farmer organizations, and the inability to take risks constrain their ability to adopt alternative practices. In North America, where the challenges to farming are of a different order, farming systems research can still bring its strengths and insights to bear on a number of different issues, including identifying the limitations to the adoption of sustainable agricultural practices. However, an appreciation of the nature of environmental, agricultural, and social systems, and of the multiple criteria of sustainability, is fundamental.

Frameworks for Sustainability Assessment and Evaluation

Among the core objectives of the FSR project is assessment of specified elements of farming systems, prediction of the implications of change both within systems and beyond their boundaries, and evaluation of these conditions and possible responses in the context of the goal of sustainability. Such work draws naturally on a range of research traditions and approaches. A variety of general and conceptual frameworks have been designed to enable analysts to assess sustainability in other types of managed and unmanaged ecosystems. Several of these are reviewed in greater detail elsewhere (Smithers 1997), but a few are noted here for illustrative purposes.

Examples of conceptual frameworks for assessment include the Pressure-State-Response model, emphasizing system processes (Rapport and Friend

1979); a Compliance-Diagnosis-Warning typology, emphasizing system conformity to standards (Cairns et al. 1992); and criteria of Vigour-Organization-Resilience, invoking certain ecological properties of systems (Costanza et al. 1992). Also, Smit and colleagues (1997) and Gallopin (1995) have attempted to characterize and assess sustainability in agriculture in the concepts, language, and methods of health. Most recently, MacGregor and McCrae (2000), among others in the Environment Bureau of Agriculture and Agri-food Canada (AAFC), have used a Driving Force–Outcome–Response model to understand how forces influencing agricultural activities, including new technologies, create environmental consequences engendering societal responses, comments about which appear in Chapter 2. While the usefulness of this approach and its methodological contributions are still unresolved, it does highlight the important role of information (indicators) for purposes of evaluation, and the goal of strategic intervention to address diagnosed threats to sustainability or health.

Working from the perspectives of agroecosystem analysis, some analysts have employed ecological network modelling approaches to assess system performance (Dalsgaard et al. 1995). Tools such as these are appropriate, and necessary, to assess in an iterative fashion the implications of different types and combinations of farm management practices for certain attributes of the farm system. To the extent that component models are linked via a macroscopic model, such approaches also have the ability to account for changes throughout whole farming systems.

Elsewhere, Stockle and colleagues (1994) proposed a framework that accommodates the use of a weighted index of farming system sustainability based on combined adjusted scores pertaining to selected system attributes and perceived constraints to sustainability. While the method does not account for linkages and interdependence within the system, it does provide a summative evaluation of the relative sustainability of a farming system. Dumanski and colleagues (1992) employ the concept of "framework as pathway" leading to a systematic approach for assessing the prospects of sustainable land management systems. This framework is underpinned by certain specified preconditions for sustainability and moves systematically through system description, problem/constraint identification, and the application of diagnostic criteria and performance indicators (see Chapter 5). This results in an assessment of the likelihood of sustainability. A strength of this method is its explicit identification of a set of criteria upon which the remainder of the framework is built and against which the system is judged.

The present framework for assessment and problem solving is distinct from other sustainability initiatives for at least two reasons. One is the central role played by the clear identification of a problem or set of problems located in a specific region and/or for a particular farming community. As a result of this, the conceptual and abstract aspects of sustainability form the context rather than the focus of research efforts.

A second distinction is the overt use of the classic farming systems approach, including the designation of different "targeting" domains (Wotowiec and Hildebrand 1988; Moore 1995). For the FSR framework, they range from the conceptual, where elements of sustainability and farming systems are described and form the theoretical rationale for evaluation, to the consultative, where input from the farm community is sought to determine problematic issues in sustainable agriculture. Based on the specification of a particular issue, the analytic context follows and involves the choice of indicators, their measurement, and computation in appropriate models. Subsequent to analysis is the evaluation domain, where results are generated that address the initial sustainability issue/problem and supply useful information for farm decisions and policymaking.

Clearly, each of these domains requires further articulation in its own right. For example, work in the analytic domain involves the development and application of diagnostic indicators and formal analytic structures for their manipulation and assessment. One potentially useful strategy involves the development and testing of certain indicators for agreed-upon assessment endpoints, such as quality of life, soil and/or water quality, economic viability, and so on, for which a wide variety of applications is anticipated. Although quality of life is not addressed by McRae et al. (2000), this approach is consistent with the Driving Force–Outcome–Response approach of AAFC. A variation of it, as described in Chapter 6, looks explicitly at linkages among sustainability indicators. A discussion of how to study these sustainability linkages is presented in Chapter 5. For this chapter, however, the focus is on the research approach and on the selection and sequencing of components for effective consultation, analysis, communication, and adjustment.

The proposed framework (Figure 4.1) uses a standard farming systems model that can be adapted to a particular case. Employing a systems approach to farming encourages the kind of thinking that is needed to understand sustainability problems, namely, interpretations that are holistic, hierarchical, and dynamic (Waltner-Toews 1996).

Conceptual Domain

In the first stage of the FSR project and research framework, basic definitions and assumptions are sorted out and agreed upon to provide the rationale and structure for understanding particular problems. The logic and effectiveness of the assessment and problem-solving exercise depend on establishing the researchers' ability to link specific details and findings to a broader set of concepts and ideas that provide the means for communication with stakeholders and other interest groups. The three elements highlighted at this stage concern the definition of criteria for sustainability, a

Figure 4.1

A framework for problem solving in sustainable farming systems research and extension

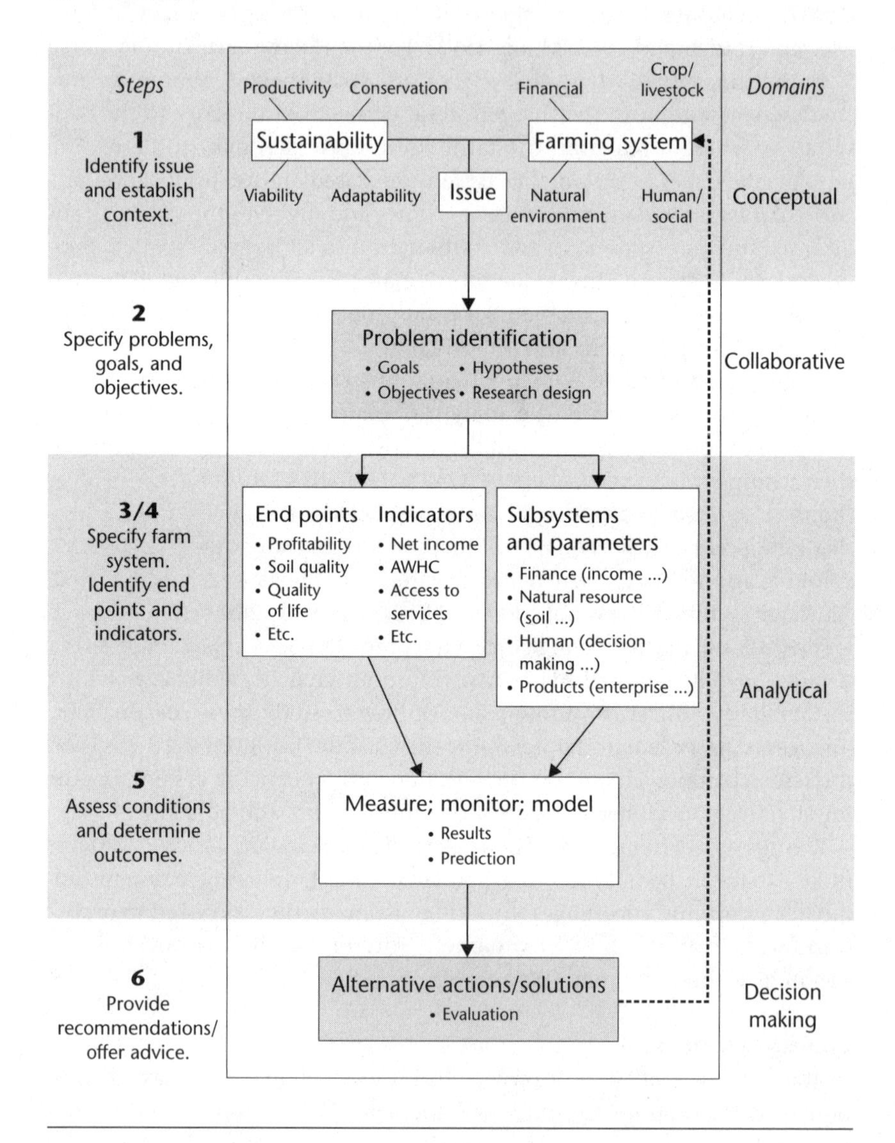

generalized understanding of farming systems, and the identification and close specification of relevant and researchable issues or problems.

Defining Sustainable Agriculture

According to Harrington (1992), sufficiently comprehensive definitions of sustainable agriculture will include ecological, social, economic, and ethical dimensions. The FSR project's working definition of sustainability reflects these qualities. Sustainable agriculture is understood to refer to production systems that are environmentally benign (or enhancing), economically viable, and socially acceptable.

Another approach to defining sustainable agriculture is to delineate a number of key elements that capture the spirit of sustainability (or health) and in turn can be treated as both sustainability criteria and/or goals. Based on reviews of ecosystem health or integrity (Kay and Schneider 1992; Rapport 1995; Woodley et al. 1993), agroecosystem sustainability (Altieri et al. 1983; Conway 1987; Dalsgaard et al. 1995; Gallopin 1995; Marten 1988), and other sustainability literature (Bryden 1994a; World Commission on Environment and Development 1987), five such system-level attributes are suggested as forming the basic elements of sustainability: productivity, conservation, viability, and adaptability.

Productivity refers to returns over input costs; security is understood here to be community viability (especially as the consequence of the relevant farming systems); protection is the cost to the community of any conservation remedial actions that may be necessary; viability is a measure of farm-family income; and the quality of life of local community members, including the farmers, is what is referred to as acceptability. While it would be possible to select other such properties (and in different cultural and technological contexts others might be appropriate), this set captures a useful set of integrating (between subsystems) themes that transcend subsystems and span a variety of recently expressed threats and issues in sustainable agriculture.

As noted above, such properties are hardly unrelated – and debate over their meaning and potential relevance invariably has them converging with each other and with the overriding notion of sustainability. Like definitions of sustainable agriculture, these components are intentionally not expressed in precise (system/subsystem-specific) terms because they reflect values and goals that have general applicability in society, the cultural and institutional context for that agricultural system. When the terms are broad, they are also ideally suited to referring to both the socio-economic and non-human or biophysical elements of farming systems, thereby enhancing integrated farming system analysis. For instance, productivity as a criterion for sustainability can be applied not only to the services available from natural ecosystems in the farming system but also to both the livestock and/or cropping systems. The concept of productivity is also applicable to the human/social resource dimension of farming systems in terms of how well individuals and social groups involved in farming activity can fulfill their roles.

Farming System Components and Interactions

Also significant at the conceptual level of the assessment framework is the definition of "farming systems." For the most part, farming systems can be defined in terms of an overall approach to farming, reflecting goals, abilities, resources, and circumstances pertinent to the farm operation – all of which result in decisions concerning the composition and operation of these human-made systems (e.g., Ikerd 1993). While farming systems are largely defined by their main inputs and products (cultivation, crop selection, pest management, land allocation, etc.), the focal point for these systems is often viewed as the farm operator as decision maker.

A key feature of the internal components of any farming system is their diverse nature, a quality that may be hidden in the versatility of farm operators but is patently obvious when outside experts are called in to aid in analysis. Thus, any research team investigating sustainable agriculture issues, in particular farming systems, will have to include proficiency in the theory and methods of a number of disciplines. Expertise in the concepts and methods of animal and crop science, agricultural economics, engineering, ecology, geography, land resource science, rural extension, and sociology *can be combined* to foster a comprehensive and inclusive approach to both the conceptualization of systems and problems and to their analysis.

A principal concern in sustainability assessment is to ascertain the basis for decisions through an understanding of both the elements of the farm system (about which decisions are made) and the social/cultural context that motivates and directs farm-level decision making. Further, in keeping with a focus on integration and feedback within systems, there is (or should be) interest in the recursive nature of these relationships, recognizing that management decisions have consequences at both the farm scale and beyond. Farming systems are not simply a sum of discrete components; they are complex and highly interactive, with significant feedbacks.

Merging concerns about sustainability with farming system components has the potential to generate a wide-ranging set of issues, many of which are interrelated. Among the most abiding of agricultural sustainability-related issues in Ontario and throughout North America is the desire to ensure the conservation of natural features and resources while enhancing and improving the viability of farming systems in the same region. Although these can be viewed as mutually exclusive, in some areas progress is being made in achieving a satisfactory balance (OFEC 1994).

Problem Identification

The identification of a particular problem or threat to sustainability focuses efforts on dealing with the practicalities and key attributes of a specific difficulty. The importance of such an exercise is well recognized in applied assessment research where issues of social/political values and multiple interests apply. For example, a defining feature of most environmental impact assessment frameworks is the process of "scoping," where valued environmental components and associated issues are defined and carried forward into the assessment process.

The same logic holds here. In North America, particular sustainable agriculture concerns are being raised by a number of groups and individuals, including the state, commodity groups, farm operators and their families, representative farming organizations, consumer/community groups, and members of the public. In most cases, sustainability issues arise because there has been direct experience with a significant problem (such as poor returns from crop and/or livestock production) or there is the potential for one to develop (such as the loss of a valuable wetland). Another approach to specifying particular sustainability problems is to rephrase them in terms of constraints to sustainability (Stockle et al. 1994). Thus, in the example cited earlier, the concern with poor returns from crop or livestock production can be considered in light of the conditions that are leading to poor productivity. Adopting this perspective emphasizes the need to address underlying or root causes of problems and encourages those involved not to accept short-term or superficial solutions.

The range of sustainability-related issues affecting agriculture is well documented and includes a host of social, economic, and environmental factors and processes. Among the more commonly voiced concerns in the southern Ontario agricultural system is the potentially negative consequences for human and environmental health from excess nitrogen and phosphorus, as well as other pollutants, in groundwater (Goss and Barry 1995). FSR at the University of Guelph has been directly involved in several initiatives aimed at addressing these problems. An example from recent consultations with resource management agencies and agricultural sector partners serves to illustrate both the opportunity and the protocol for issue-driven research collaboration.

Current concern for water quality in a local, largely agricultural subwatershed has provided an impetus for discussions on a potential collaboration between FSR, a local resource management agency, the local farming community, and a downstream urban municipality to improve water quality in the region. At this point, hypotheses and research designs can be developed in collaboration with interested parties and in the context of their goals and objectives. Waltner-Toews and Wall (1997) note that for agricultural sustainability (or health) issues, it is useful to recognize that two kinds of goals are at work: primary and operative. The former refer to those ideal aspirations that act as overall long-term objectives, while operative goals are subordinate and considered the means to achieving primary goals. With respect to the FSR example, the primary goal is to improve water quality in the watershed while the operative goals are more specific and include (among other things) correcting manure management practices, modifying fertilizer handling, and protecting riparian areas.

Distinguishing goals in this manner acknowledges issues of scale and hierarchy, a distinguishing feature of farming systems research. Primary goals are clearly at a broader scale than the operative ones; the former encompass the latter. Specifying goals is also useful for accommodating the potential tension between "science and farmer participation," recently articulated by Caldwell and Christian (1996). In the southern Ontario example, the primary aim of improving water quality is one to which all interests can readily agree. When the focus moves down the scale to more specific operative goals and/or objectives, a logical division arises with certain farm operators (e.g., livestock producers) and certain farming system experts (e.g., land resource scientists) taking on specific projects motivated by particular objectives. Although conflicts might arise at this level between the interests of researchers and farm operators (in that each may resent and resist changes suggested by the other), it is possible for such differences to be resolved in light of their common attachment to the primary goal.

Analytical Domain

When the problem has been clearly identified and communicated, and the goals, hypotheses, and research design have been determined, the next stage of the research and of the framework involves more formal integrated analysis. Here the concepts, theories, and methods of the natural, agricultural, and social sciences are brought to bear on now well defined issues and tasks, and on a conceptual understanding of the farming system. It is at this stage that effective interdisciplinarity is most critical and its challenges most visible. The key challenge is to determine those components of the farm system that are salient for the task at hand, and to select and apply paradigms and methods that are appropriate for that (usually multidimensional) problem. The former effectively represents a scientific "scoping" of the research task. The latter demands an appreciation of the potential contributions and limitations of methods in the natural and social sciences. The desired end is an integrated research strategy in which the environmental, economic, and social dimensions of sustainable agriculture problems are considered in a systems perspective. The specific mix of approaches (environmental systems modelling for purposes of prediction, behavioural research to explain farm decision making, statistical analysis and validation of proposed indicators, and so on) and the extent to which they are fully (and physically) linked or operate in a parallel and complementary fashion will vary in each application.

Specifying the Farming System

Farming systems consist of subsystems at smaller scales of aggregation, and exist as part of larger systems in a hierarchical fashion (Allen and Starr 1982; Izacs and Swift 1994; Weber 1996). According to Bird and colleagues (1984), the first step in a systems approach is to identify the boundaries (and the elements) of the system under consideration, which leads to defining those components of the farming system that are internal to this realm and those that lie outside it. The former include crop/livestock systems, farm economics, human interactions, and on-farm ecosystems that can be isolated and examined in terms of how each one influences and is influenced by the others. By contrast, systems pictured as external to the farming system refer to such elements as markets, policy, climate, and natural resources, all of which are of particular interest for sustainability issues.

The designation of a particular problem will implicitly highlight one of the subsystems within the farming system. The southern Ontario example cited above would begin with a focus on crop and/or livestock subsystems. However, the selection of key initial subsystems does not suggest that the research is artificially limited to these aspects of the farming system alone. Quite the opposite. Other subsystems of the farm and the local agricultural system are dealt with in time and in terms of their relationship to aspects of the initially relevant subsystems and the broader sustainability issue. Thus, if and when barriers to adoption of better manure management are studied, conditions in both the financial and human/social aspects of the farming system (as well as a number of external factors) would become relevant.

Choosing Measurement Endpoints and Indicators

Essential to the sustainability analysis is the selection of appropriate indicators that will depend on a number of factors, including the designated problem, scale, type of farming system (such as livestock, cash crop, horticulture, and so on), and specific component of the farming system implicated in the sustainability issue. Thus it may be that attention is directed at all the farming systems in a region where the problem identified concerns soil erosion; it may be all livestock operations in the case of concern over nitrate loading in a damaged watershed; or it may be one highly diversified industrial farming system that is perceived as problematic because it is threatening the sustainability of neighbouring smaller farming systems.

Also important for indicator selection are the significant endpoints chosen; they will have a direct connection to the goals and objectives that are driving the research (Cairns et al. 1992). In the case of the southern Ontario watershed, for instance, a target parts per million measure of nitrogen (N) and phosphorus (P) will be decided on as a desirable endpoint for the watershed. Various changes in farming practices can then be monitored with respect to how they affect the N and P targets.

Regardless of which endpoints are chosen, all significant variables need to be operationalized into a number of measurable indicators. Because there has been significant interest in the development and testing of agri-environmental indicators of sustainability and health (e.g., Agriculture Canada 1993; Smit et al. 1997; Benbrook and Mallinckrodt 1994; Eswaran et al. 1993), there is a wide array of choices for sustainability analysts. Earlier indicator efforts could be described as relatively conceptual, involving a search for particular indicators in light of general sustainability problems. The integrated framework presented here somewhat reverses the emphasis and suggests instead that the more productive search is for particular sustainability problems that can be addressed with generally accepted indicators. For instance, soil quality can be reflected with specific values for nutrient balances, water-holding capacity, or percentage of organic matter. The AAFC Environment Bureau assessed soil quality based on the use of such indicators as the degree of risk of water, wind, and tillage erosion, soil organic carbon, soil compaction, and soil salinization. Likewise, quality-of-life indicators include both "objective" measures such as net income and "subjective" assessments related to various perceptions regarding the availability of services, conditions in farming and the rural community, and future outlook (Filson 1997). Earlier indicator efforts could be described as relatively conceptual, involving a search for particular indicators in light of general sustainability problems. The integrated framework presented here somewhat reverses the emphasis and suggests instead that the more productive search is for particular sustainability problems that can be addressed with generally accepted indicators.

Empirical analysis has also documented ranges for a number of measures that will affect soil and water quality, agricultural economic viability, and human physical and social health. What science cannot do, however, is dictate the acceptability of these impacts. Such decisions are related to the values belonging to those who enlist the help of farming systems research for dealing with specific sustainability issues.

Measuring, Monitoring, Modelling

A fifth step in the problem-solving framework is directed at assessing conditions and determining outcomes based on the nature of the data gathered and the approach selected for their manipulation and analysis. Opportunities for explanation and prediction arise from the use of rigorous analytical methods based on inferential statistical analyses, simulations, and optimization techniques. Of special interest are mathematical modelling techniques for the analysis of agricultural systems. At this analytical stage, individual researchers, working within accepted methodologies appropriate to their disciplinary interests and roles, not only generate findings specific to selected issues (e.g., developing and testing indicators of quality of life) but also contribute information to a more integrated venture. Current work focuses on the development of functional optimization and simulation frameworks where trade-offs and interactions among the components can be assessed systematically and iteratively (Goss 1994; Yiridoe and Weersink 1997). The use of such models for addressing sustainable agriculture issues overcomes some of the limitations associated with farming systems research and extension work. As it has been practised for most of its history, farming systems analysis has precluded opportunities for long- (or even medium-) term experimentation as well as for transferring findings to conditions in other locations (Harrington 1992). With the adoption of modelling techniques, it is possible to extend both the temporal and a spatial component in the analysis, thereby increasing the potential information available for evaluation.

Evaluation and Decision Making

Integrated farming systems research represents an example of what Waltner-Toews (1996) has termed "mandated science." The clear intent is to offer information and advice as a basis for improved, or at least informed, decision making at the farm scale and beyond. Hence, from the analytic exercise described above, there is need for the broader question "So what for sustainability?" What are the implications of findings from an integrated analysis for the problem initially described (recall the water quality example earlier) and, just as important, for the sustainability of regional farming systems?

So while Figure 4.1 suggests a terminal point in the proposed problem-solving framework devoted to discussion of alternative actions and solutions, it is in many ways yet another beginning. Recognition of the interconnectedness of systems begs further questions. Solution options for one issue must be considered relative to their implications elsewhere in the system. In effect, there are not only responses to issues but also potential issues from responses. Systems-oriented research must accommodate not only first-order outcomes associated with certain courses of actions but also the recursive nature of systems. Bounding of the research scope can and should occur in consultation with sector partners, but should be underpinned by a commitment to understand the systemic aspects of farm-level sustainability and associated actions towards that end. Adequate assessment of the prospects for sustainability in farming systems requires a system-level perspective. While the approach outlined here is driven by locally (and perhaps narrowly) defined concerns that often focus on selected issues and subsystems, concern for the sustainability of whole-farm systems and of farming regions demands appreciation of interactions within and between those systems, and of the implications of change, socially, economically, and environmentally.

Conclusions

This chapter reviews the sustainability problem-solving framework being used by the FSR team at the University of Guelph, which has addressed important issues in both farming systems analysis and sustainable agriculture. The framework, which is grounded on the premise that North American farm operators will be able to adopt more sustainable practices only when they have clear standards against which to gauge their success or failure, has been presented as an integrated and practical method that focuses on solving problems in sustainable agriculture.

The framework is integrated in several ways. First, and in the most general terms, it draws together traditional farming systems methods and sustainable agriculture issues. Once a specific problem has been articulated, the system functioning can be analyzed and results implemented in improved techniques. A second integration feature involves the multidisciplinary approach that the FSR framework embodies. Although formal scientific analyses may be confined to disciplinary fields, they become synthesized when solutions are generated in a latter phase and in the context of a primary goal. This process requires integration across disciplines and between the realms of public interest and scientific expertise. In the FSR framework, neither can operate effectively without the other. Equally possible, and highly desirable, is the development of more formally integrated models that capture the linkages and interactions between the environmental, economic, and social dimensions of farming systems and facilitate whole system-level prediction. The development of such tools is a needed component and a valuable by-product of systems evaluation research.

Sustainability has not been, is not, and never will be a fait accompli. As values, goals, and objectives evolve within the socio-cultural context that informs agricultural practices, so too will the identification of problems and the solutions associated with them. The research framework described in this chapter offers a structure that accommodates and incorporates different needs, perspectives, and uses, and has been presented here as a viable and practical approach in applied farming systems research in support of sustainable agricultural systems.

5
Developing Predictive and Summative Indicators to Model Farming Systems Components

Michael J. Goss, John R. Ogilvie, Glen C. Filson, Dean A. Barry, and Santiago Olmos

In farming systems research, socio-economic as well as biophysical components receive the attention of researchers from a wide range of disciplines. However, these components are often studied separately and their interactions are either ignored or not well understood. The main objective of this chapter is to examine the linkages or interactions that exist among the various components of farming systems and to suggest an approach to farming systems research that integrates social and biophysical dimensions.

Farming systems research has often been used in developing countries to identify the roles of individuals in primary production and in meeting the needs of the family for food, fuel, and fibre. Application of such an approach in developed countries has often focused on gender issues or the integration of information on the socio-economic aspects of adoption of new or alternative practices. A further challenge to farming systems research has been to develop an understanding of the needs for a sustainable agriculture that can support the growing human population. In countries with a highly developed and intensive agriculture, the need is to integrate societal needs, producer needs, land capability, and environmental protection into a predictive framework.

Sustainable Agriculture and the Use of Indicators

The need to promote and attain sustainable development is a result of a growing pressure on natural resources at a global scale. Soil erosion, increased use of pharmaceuticals and pesticides, contamination of drinking water, and deforestation are among a multitude of problems that have been caused by ever greater demands for agricultural products, and have prompted many to question the environmental gains made by the application of a reductionist approach to research on production methods. The emerging paradigm of sustainable agriculture emphasizes the need for a holistic view of the countryside, with agriculture being indefinitely productive and supportive of life. In a sustainable farming system, "social considerations are balanced with environmental and economic concerns" (Ikerd 1993, 152). According to Ikerd (1993, 151), such an agriculture "must use farming systems that conserve resources, protect the environment, produce efficiently, compete commercially, and enhance the quality of life for farmers and society overall." Dumanski and Smyth (1994) listed five objectives of sustainable land management:

- to maintain or enhance production and services (productivity)
- to reduce the level of production risk (security)
- to protect the potential of natural resources and prevent degradation of soil and water quality (protection)
- to be economically viable (viability)
- to be socially acceptable (acceptability).

At the University of Guelph, the Ontario Agricultural College listed seven criteria for sustainable agriculture:

- production of safe and nutritious food
- maintenance of the soil base
- protection of water resources
- protection of air and atmosphere
- conservation of nonrenewable resources
- maintenance of economic viability
- social acceptability.

There is an obvious similarity between these three descriptions, which have been constructed by individuals viewing sustainable management from a variety of vantage points. Each contains elements that have political as well as scientific overtones. However, there have been criticisms of the application of concepts associated with sustainable development to modern farming. Agriculture has often been subject to regulations that differed markedly from those applied to other industries, making the governing economic framework difficult to establish. Aside from the ethical, political, economic, and biophysical issues associated with the concept of sustainable agriculture, one important source of complexity is the need to assess or measure progress towards the achievement of this goal. The complexity of the agri-food industry means that not every process and component can be adequately monitored, so indicator parameters have to be selected that act as surrogates for functions or properties. What indicators could or should be used? What and whose values would be included in developing such indicators? At what scale, and how frequently should assessments be made?

Goss (1994) argued that the similarity in the definitions surrounding sustainable agriculture permitted the development of indicators that had general application. He defined diagnostic indicators as "measurements of inherent or dynamic characteristics of a farming system." They represented the minimum common set of measurements necessary for the evaluation of sustainability. Goss (1994) also developed a system of bounded indicators, defined as "indicators with set limits, above or below which the properties projected by predictive calculation would cause the system to fail the test of sustainability."

In a systems approach, the relationships between different parts of a system (rather than the performance in one component of the system) are regarded as most crucial. These relationships include the interaction of social, economic, and environmental components, as well as the interaction between human and nonhuman elements of the biosphere. Thus, a holistic view of the qualities of the whole system (rather than just economic performance, for example) is characteristic of sustainable agriculture. Measuring progress towards sustainability must therefore imply some sort of integration of the different elements of a system. An indicator of water quality, for instance, would tell little about the sustainability of a system if we do not take into consideration the costs to the farming community or other parts of the system.

For communities to improve their performance in striving towards the goal of sustainable production systems, the concept itself must be well understood. Only then can suitable indicators of sustainability be selected. Furthermore, the general public must be well informed about the concept and the chosen indicators if they are to evaluate the progress of their community. In other words, it is important that not only researchers but also government officials, farmers, and the public are able to determine whether practices in their community are sustainable.

The idea that an agricultural system is sustainable if it provides commodities (food, fuel, and fibre) or income and is supportive of societal goals must be an important facet of discussions about indicators. Indicators of the biophysical aspects of a farming system can be selected that are consistent with the goals of sustainable production. Using the soil subsystem as an example, the potential areas for consideration can be identified. The essential factor that determines the effectiveness of a soil in agriculture is its condition as a support for plant growth and development. Soil conditions may be assessed under paradigms of quality or health. The quality of soil "relates to the capability of the soil for production or provision of other services beneficial to humans such as pollution attenuation" (McCullum et al. 1995, 6). Indicators that have been proposed for describing soil quality include measures that are site-specific, such as pH levels, texture, and organic matter content. They often come from the minimum data sets necessary as inputs to simulation models of crop growth (Goss 1994). Those indicators provide an idea of the ability of a soil to meet certain management objectives. It must be made clear that a soil that is of satisfactory quality for corn production may not be suitable for other crops. Therefore, the indicators used for assessing soil quality must be based on system objectives. The term "soil health" equates soil conditions with the health of an organism. In this paradigm, soil conditions include "factors which may be unrelated to the achievement of management objectives" (McCullum et al. 1995, 6).

Although it has been suggested that to monitor the health of a soil system, suites of indicators are needed because "the development of absolute standards for assessment of ecosystem health may not be scientifically possible" (McCullum et al. 1995, 11), the same argument can be made for developing a suite of indicators using a quality paradigm. In the case of the soil subsystem, Goss (1994) proposed that bounded indicators could be developed for the maintenance of the soil base by limiting erosion to the rate of topsoil production, preventing deterioration of soil structure, and limiting the application of toxic metals or potential biocides. These indicators would measure aspects that relate to both crop productivity and ecology. They reflect the needs of both society and the producer, and they can be influenced by readily available management practices.

Whether indicators of the socio-economic aspects of a farming system can be selected that also conform to the goals of sustainable production is worthy of closer examination. Using the list of objectives for sustainable land management, some potential indicators can be linked to given objectives (Table 5.1).

In these examples of potential socio-economic indicators, some operate at the level of the individual farm while others deal with the community. Indicators of agricultural system sustainability can be either summative or predictive. Summative indicators are thought to be those the community applies to entities within its purview. Predictive indicators, on the other hand, are those for which valid measurable data can be obtained at the farm level. Examples of the two types of indicators are included in Table 5.2.

Predictive indicators represent potential influences of the farm on the community (see Chapter 3 for impacts of intensive agriculture on rural communities). These influences are seen by their effect on the summative indicators.

Table 5.1

Socio-economic sustainability objectives and possible indicators

Objective within sustainability	Possible socio-economic indicator
Productivity	Return over input costs
Security	Community viability
Protection	Cost to community of any remedial actions
Viability	Farm-family income
Acceptability	Quality of life of local community members

Table 5.2

Summative and predictive indicators

Summative indicators	Predictive indicators
Water quality	Crop yields relative to potential
Bacterial concentration	Nitrogen-use efficiency
Stream phosphorus	NH_3 losses from barns
Stream biochemical oxygen demand (BOD) load	Surplus phosphorus load
Community family income	Return over input costs
Ecosystem health	Off-farm work
Quality of life	Farm-family income
Community viability	Contribution to community income
Value of milk quota	Debt-to-income ratio

Linkages between the farm and community can be thought of as the cause-effect or statistical relationship between the predictive and summative indicators. For evaluating the impact of the surrounding community on farm viability, indicators in the community could be predictive and those on the farm considered summative. For example, the debt-to-income ratio might be a predictive indicator of viability for a community, whereas the value of the milk quota might be a summative indicator of the effectiveness of supply management at the farm level.

The extent of a community's "social capital" can be a useful summative indicator. Smit and colleagues (1997), for example, have emphasized the importance of support networks in understanding community well-being. They have then introduced the concept of social capital, which is an attribute of community well-being: "Indicators that reflect the stability of the community include rate of farm turnover, number of young farmers, and availability of off-farm work." Political capital can also be assessed from the level of political participation and degree of environmental regulation. Similarly, indicators of economic viability might be the level of net farm income or average household income for the community.

Quality of life may also be useful as a summative indicator, but one that provides feedback to the level of a predictive indicator. Chapter 9 shows how assessments of farmers' quality of life may be made using people's perceptions of their well-being as a function of other features of their lives. McCoy and Filson (1996) used this method to discover that off-farm work often affects western Ontario farmers' perceived quality of life negatively. Respondents who work off the farm were less satisfied with the quality of time spent with spouse, children, and friends and personal time. Richmond and colleagues (2000) employed the same approach in assessing nonfarm rural people's perceived quality of life in the North Durham Region of

Ontario. Income, education, access to community services, and numbers of children in the home were important factors affecting these people's quality of life.

It is important to recognize that a full assessment of sustainable farming systems requires indicators dealing with social, economic, and biophysical aspects, and that the three aspects may have to be considered simultaneously. For example, a survey of dairy farmers and their families in the Grand River Watershed indicated that a supply management system for milk was regarded as a crucial operating condition (Chapter 9, Table 9.5). The importance of the quota/supply management system to the Grand River dairy farmers may indicate that the farming system is extremely dependent on economic support. Such a system is considered by some to be subsidization, although many would say that it is not. If supply management is a subsidy, it may be perceived as "unhealthy" (Rapport 1994). Other authors (Smith and Saunders 1995; Bradshaw and Smit 1997) have indicated that relatively recent reforms in New Zealand's agricultural sector, which involved reductions in subsidies, led to deterioration of the farm economy as well as landscape features.

It appears that the economic strength of farm enterprises and the rural environment are closely connected. Furthermore, the economic support received through supply management may contribute to the implementation of sustainable practices at certain levels of a farming system. Thus, treating economic support given to farmers as an indication of economic ill-health is to take too narrow a view, especially if the benefits to the biophysical components of farming systems are considered.

The following variables could be monitored, and the relationship among these variables can be more clearly understood after monitoring is done in different settings and under different conditions:

- off-farm work participation (by farm operator, spouse, children)
- importance of off-farm earnings to total family income
- debt/asset ratio of farm enterprises
- importance attached to economic support (e.g., quota system) by farming community possibly indicative of level of (in)security
- level of economic support given to farmers (expressed in terms of cents per kg of farm output)
- impact of changes in farm returns on conservation practices
- level of education of farming population
- percentage of farmers who have enterprise expansion plans
- percentage of farmers who have no successor(s)
- percentage of farmers who intend to invest in nature conservation
- levels of income among members of farming community

- levels of income in the rural community
- off-farm income/farm income ratio
- "standard" indicators of economic activity
- access to health services
- nitrogen-phosphorus-potassium efficiency
- on-farm emissions of ammonia and other gases.

Most of the indicators listed above could be categorized as predictive indicators. Moreover, the list is not exhaustive as it consists mainly of indicators used to monitor the farming system's economic and social health. A number of other indicators may have an effect on summative indicators (for example, quality of life). Understanding the relationship between the predictive indicators and the summative indicators of a farming system such as that described above is crucial for sound policymaking.

A model that includes the biophysical, social, and economic aspects of a farming system must also include a farm-family decision-making component if it is to be useful for policy analysis. The farm-family decision-making component would most likely be a rule-based expert system connected to biophysical, economic, and social modules. Decision making by the farm's manager often includes discussions with family members and off-farm interactions (Dent et al. 1995). A farm-family classification scheme would be required to predict the response of different family types to policy and other changes in the on- and off-farm environment (Ritchie and Dent 1994). Because of the potential complexity of farming system models, it is important to include only those components and relationships that are most important for understanding the system. Assigning the level of importance to different components and relationships within the system is complex, however.

Several factors would influence a model of farm-household decision making. A list of external and internal influences on the decision to increase, decrease, or maintain the amount of resources engaged in agriculture was developed from a survey of 7,000 farms in western Europe (Bryden 1994). Farms were classified as engagers, disengagers, or stable, depending on the change between 1981 and 1991 in the amount of resources engaged in agriculture. The main influences from farm and household characteristics on farm adjustment appeared to be:

- proportion of income from different sources (farming, off-farm, on-farm non-agricultural, social transfer payments)
- size of the farm business
- number of economically active persons in the household
- age of the farmer
- labour input into farming by farmer and household

- amount of off-farm work by farmer's spouse and family
- background of farmer and spouse
- educational level of farmer and spouse.

The main external influences were:

- nature of the labour market, especially the nature of non-agricultural work
- nature and extent of non-agricultural opportunities on the farm
- small versus large-scale agricultural structures (more pressure to seek non-agricultural income or exit agriculture if farm is small)
- implementation of agricultural and rural policy.

These various influences on decision making could be used to adapt a behavioural model framework to different circumstances. The farm-household decision-making model must take account of the fact that farmers may have multiple goals. This approach for incorporating household decision making into a farming system model was described as the psychological-behaviourist approach by Edward-Jones and McGregor (1994). They point out that it has been applied to farmer perceptions of their quality of life, farming values, satisfaction with farming, and risk taking, but with little experimental evaluation. It is suggested that this approach, combined with economic goals, will probably be the most successful approach for modelling farm-household decision making.

Time-Dependence of Assessments

The structure of a soil can change dramatically over a short time period, such as the period of passage of a wheel or plough. Trends can be identified by monitoring at the same season, and at the same time in the cycle of crop development. Some authors argue that it is those properties that change at such "intermediate" time scales that have the most potential as indicators of soil conditions and, by extension, of sustainability.

Because sustainability implies a long-term condition, evaluation of a system also needs to consider changes with time. The components of the system that need to be evaluated and that generate values for the various indicators will interact with each other to give rise to the system behaviour over time. A means of linking the economic, social, and biophysical components to form a system model should reflect our best understanding of how the components interact, so that outputs from the model can help us understand overall system behaviour.

One approach for linking various system components to understand their dynamic behaviour over time is the System Dynamics method developed by Forrester (1968) for industrial systems. It assumes that a system has a goal,

Figure 5.1

Examples of negative and positive feedback loops according to the System Dynamics method

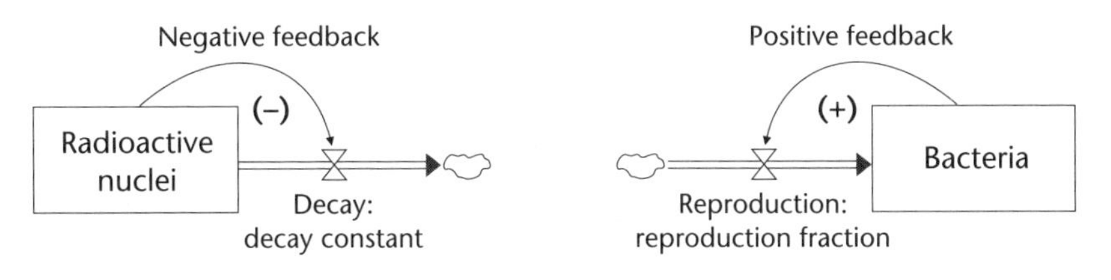

Source: Modified from Martin 1997.

and its behaviour over time to reach that goal is controlled by feedback loops. The feedback loops are an information flow that represents the input for controlling the rate of change of the various model component outputs. Information about the level of an output flows back to control the rate of change of that and perhaps other outputs. The System Dynamics approach therefore relies mainly on two types of variables: level or stock, and rate or flow. The stock variables are connected to the flow variables through feedback loops (Figure 5.1).

Negative feedback systems exhibit goal-seeking behaviour, which may be asymptotic in simple systems or oscillatory in more complex systems. An example of a negative feedback system is room temperature controlled by a thermostat operating a furnace that heats the room. Another example is decay of radioactive nuclei, because an increase in the number of nuclei increases their rate of loss (Figure 5.1). Positive feedback causes growth and change. An example is the increasing growth rate of bacteria in a flask as their population increases (Figure 5.1). The two types of feedback combine to give all the behaviour observed in systems (Martin 1997).

In this approach, there is no clear distinction between cause and effect because a stock variable feeds back to control the flow variable that controls the stock variable. The advantage of using this method of constructing component linkages for a system is that it allows the various economic, social, and biophysical components to be connected by information flow represented by the feedback loops.

Information about the levels of the various outputs of the biophysical and economic components could be inputs to community and farm-family decision-making components that could be represented as rate variables that directly or indirectly control inputs to different components of the biophysical and economic models. Determining the mechanism of the decision-making linkages would require interviews with farm and community members to understand their decision-making process.

Modelling the Linkages between Farm and Community

A major concern in rural communities is how farming systems impact the rural environment, particularly water and air quality. To predict these impacts, simulation models of the crop and livestock components of a farm need to provide outputs that can be linked to environmental impact models. To determine the viability of different farming system scenarios, models that evaluate the human resource (labour), social (quality of life), and economic aspects must also be linked to the biophysical models such as MCLONE4 (Stonehouse and Goss 1999) and DAFOSYM (Rotz et al. 1989) (Figure 5.2). Influences of the farming system and rural community on decision making by the farm family would be a further step in evaluating the long-term viability of a farming system.

The diagram in Figure 5.2 is set up to show how elements affecting the farming system can be integrated in the short term with some software currently available. The HR and Social models and the Environmental Impact models are currently ill defined. The relationship of various inputs and subsequent outputs from the animals or crops is well developed, but what the outputs do to a particular summative indicator is not well developed. The objective of our work is to use existing software to the greatest extent possible to provide solutions to problems that arise from within farming systems, and from the relationship between the farming systems and the surrounding rural community. Within the boundary of the community, there are one or more dairy and swine units and crop production units. CROPSIM, a suite of computer programs written by former FSR member Tony Hunt, can provide the outputs in addition to crop growth and yield as predictive indicators of sustainability. The existing software DAFOSYM (Dairy Forage Simulation Model from Michigan State University) is supported by a team at MSU and the University of Wisconsin. It provides output and integrates with weather data. The summative report also includes plot output that shows the probability distribution of achieving a wide range of measurable parameters over the length of weather (nineteen years in the case of Guelph weather). The existing software MCLONE (Manure, Cost, Labour, Odour, Nutrient Availability, Environmental Risk) deals with swine and dairy manure systems and manure application recommendations. This decision support system is explained in Chapter 8. Ideally, there would be flows passing from CROPSIM, DAFOSYM, and MCLONE to Process Optimization, so that this modelling technique would take account of nutrient losses from manure prior to any environmental impact models.

Although this book does not present a fully formed Human Resource or Social model, in part due to the complexities involved, the first application presented in Chapter 7 provides an example of a Whole-Farm System model to find the best ways to feed livestock and handle manure in the most economically and environmentally sustainable way feasible using mixed integer

Figure 5.2

Model components of a farming system and its links to the rural community

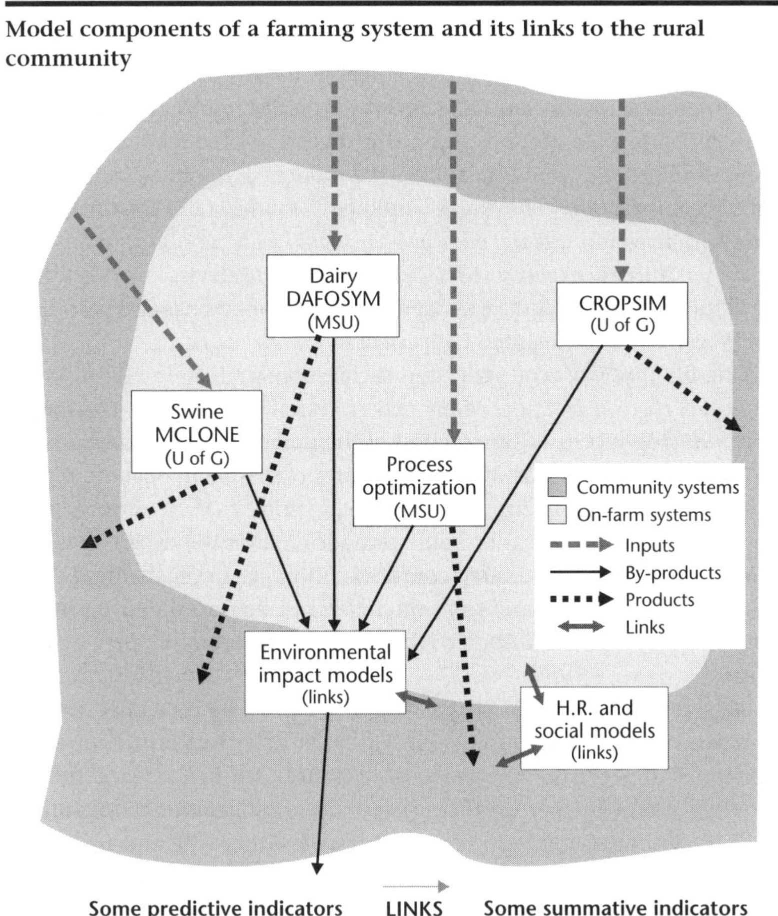

Some predictive indicators
- Soil nitrogen fllux
- Building NH_3 losses
- Surplus phosphorus load
- Off-farm work
- Return over input costs
- Community income from local farms
- Farm family income
- Farm organic loading from manure

Some summative indicators
- Water quality
- Stream phosphorus
- Stream BOD load
- Community viability
- Community family income
- Ecosystem health
- Quality of life

Abbreviations:
DAFOSYM = Daily Forage Simulation Model (at Michigan State University)
CROPSIM = Crop Simulation Model (T. Hunt, Plant Agriculture, University of Guelph)
MCLONE = Nutrient Management Computer Program (J. Ogilvie et al., University of Guelph, School of Engineering)
Process Optimization = model for balancing phosphorus flows for dairy (Michigan State University)
Source: Modified from a drawing by John Ogilvie

programming. A whole-farm model, according to Dent (1993), should ideally include farm and farm-family components (farm enterprise, farm management, ownership and off-farm work, farm-family basic needs for survival, and farm-family motivation, attitudes, and objectives) acted upon by physical and economic forces, and various factors (information, education and health, culture, kinship, peer group influencing the socio-cultural components). This decision process would most likely be modelled as a rule-based system designed using experiential data. A whole-farm model could be constructed as a shell that would call on biophysical, social, and economic submodels and data sets and the expert system (Dent 1993). Considerable work on modelling whole-farm production and environmental impacts has also been done at the Wageningen Agricultural University in the Netherlands.[1]

Linking biophysical, economic, and social submodels to the farm household would require creation of an expert system to simulate the farmer decision-making process. Construction of the rule base of the expert system would require detailed consultation with farm operators to document their decision-making process and weighting of various goals. The simulation submodels and external factors would provide data for the expert system to access in determining the desired course of action. An understanding of the feedback loops in the farm system model would help to identify leverage points where system behaviour could be most efficiently changed (Barry 1998).

To date, the FSR team has had a clearer grasp of the predictive than of summative indicators because we still lack sufficient information about summative indicators. Summative indicators, as posed here, are often those measured by agencies or groups external to the farm operation. For example, a water quality measurement, while accurately done, will sum up all the influences upstream yet may target the local farms as prime contributors. We do not know yet the extent to which the predictive indicators influence the summative indicators. This poses a major challenge for future FSR work.

The Sustainable Livelihoods (SL) approach to understanding individual and farm-household dynamics may provide a way of understanding the relationship between predictive and summative indicators in the future. Although the recent work on SL by Scoones (1998) and Ellis (2000) has potential applications for developing country farmers, it has been mainly developed for helping poor farmers in developing countries. The SL approach deserves mention here because it utilizes the traditional FSR techniques of systems thinking, integration, and holistic thinking. Livelihoods

1 F. van Evert, "Modeling whole-farm production and environment impact," Wageningen University and Research Centre, the Netherlands, online at <http://www.plant.wageningen-ur.nl/projects/modeling-framework/whole-farm/whole-farm-2003.html> (retrieved 20 July 2003).

are described as being composed of assets that include social, natural, physical, human, and financial capital, and livelihood activities as well as levels of access to conditions that social relations and institutions mediate (Ellis 2000).

Work on biophysical components has included developing or modifying models for dairy farm management, manure management, and nutrient use optimization. These models need to be linked to environmental impact models, which might involve a geographic information system (GIS) based approach. Factors identified from the quality-of-life surveys in the Grand River Watershed in Ontario provide indicators that need to be linked, probably by regression relationships, to the biophysical outputs. The quality-of-life surveys also provide information needed for weighting the various economic, social, and environmental indices that need to be summarized as a measure of sustainability of the farming system and community. FSR work is continuing at the University of Guelph to elaborate these methods for the farm-level modelling described in Chapter 6 for southwestern Ontario watersheds.

Concluding Comments

The social and biophysical dimensions of environmental issues (such as the ones that can be dealt with in farming systems research) must be integrated to make possible a more sustainable future (Woodhill and Röling 1998) Nevertheless, the integration of socio-economic and biophysical variables, as outlined above, is complicated by the fact that it is not well understood how some of these variables interact at different spatial and temporal scales. Farming systems research becomes a most challenging endeavour, as it is an attempt to understand interrelations of socio-economic and biophysical variables at different levels of aggregation. Concerns about the social and environmental aspects of sustainable development add a temporal dimension to the problems being investigated. Such concerns make it indispensable that both longitudinal and comparative studies of farming systems are made.

The distinction between predictive and summative indicators appears to be a useful tool for understanding the linkages between the farm and the community. This facilitates assessment of the sustainability of a farming system at different levels of aggregation. In the case of the Grand River Watershed dairy farming system, progress has been made in identifying farm-level indicators that have an important influence on such a crucial community-level summative indicator as quality of life, which is also an important indicator of the sustainability property of acceptability. Such relevant indicators and linkages of the effect of the supply management system for milk on dairy farmers' quality of life are discussed in greater detail in Chapter 9.

This chapter argues that in order to develop a sustainable agriculture, indicators must be used as criteria for sustainable production. These indicators can be both descriptive and diagnostic, especially when they have threshold level boundaries that take into account problems of scale. The choice of indicators must be consistent with the goals of sustainability at farm, community, and environmental levels. They must also be based on the objectives of the specific farming system. They show the effects of the interactions between the various farming system components so as to evaluate the sustainability of the system. Indicators must be selected so that they will have meaning and credibility to the people most involved with these systems: researchers, the public, government officials, and farmers.

Special attention must be paid to direct and feedback relationships (linkages) between both summative and predictive indicators. This chapter proposes a model of the main components of a farming system that integrates the indicators of sustainability and shows its links to the rural community. To be predictive, farming system models must integrate the indicators of sustainability, such as viability, productivity, environmental protection, and social acceptability, in such a way that they simultaneously consider social, economic, and biophysical indicators while taking into account farm-family decision making.

6
Modelling Farming System Linkages

Alfons Weersink, Scott Jeffrey, and David Pannell

Chapters 4 and 5 have presented a problem-solving framework for researching problems related to sustainable agriculture and examined the linkages between various components of farming systems.[1] The six-step framework described in Chapter 4 involves: (1) identifying issues, (2) specifying goals, (3) specifying farming systems, (4) identifying indicators, (5) assessing conditions and outcomes, and (6) providing recommendations. The focus in Chapter 4 was largely on the first two steps. Chapter 5 elaborated upon the linkages in farming systems, particularly on the indicators for sustainable agriculture. This concluding chapter of Part 2 focuses on the last steps of the problem-solving framework, particularly the means to empirically implement the framework.

The purpose of this chapter is to review the modelling issues faced by researchers empirically examining a farming system. The chapter moves through the framework and addresses practical issues faced by researchers at each step. However, the focus is on formulating the models for estimating the levels of sustainability indicators and the trade-offs between various indicators associated with alternative production practices. Issues discussed include the modelling approach (optimization/simulation versus econometric), unit of analysis/choice of biophysical model, aggregation, risk, and time.

Trade-off Curves

Government regulators interested in sustainability issues want to know how to balance the health of both the environment and the farm economy most efficiently. There is generally a trade-off, since improving environmental health can be achieved by reducing agricultural residuals but this likely comes at the cost of reduced producer returns as abatement efforts

1 A longer version of this chapter, entitled "Farm-level modeling for bigger issues," appeared in *Review of Agricultural Economics* 24, 1 (2002): 123-40.

are required to lower emissions. The missing information relates to the optimal level of the residual, the most profitable practices associated with that residual level, and the policies necessary to achieve the adoption of those practices.

Determining the optimal residual level (and subsequently management practices and policy) involves expressing both variables of concern, environmental and economic health, in monetary terms. Conversion of the physical impacts of agricultural practices on the environment into monetary values permits a direct comparison of the off-farm environmental costs and the on-farm abatement costs. However, such conversion often involves non-traded goods and there are inherent difficulties associated with attempting to assign values to goods that are not traded within a market. While methodological advances in nonmarket valuation techniques continue to enhance their acceptability, Smith and colleagues (1999) and Kirchoff and colleagues (1997) warn about the unreliability of benefit transfer or the use of values for enhancing environmental health from one location to another. The use of meta-analysis as a means of benefit transfer across locations is limited (Boyle et al. 1994; Smith and Huang 1995). Thus, the use of a cost-benefit framework often requires the use of nonmarket valuation techniques for each specific location and issue.

An alternative to the cost-benefit framework in quantifying the impact of agricultural production on the environment is the use of trade-off frontiers. Antle and colleagues (1998) argue that plotting economic indicators (in monetary terms) against environmental indicators (in physical terms) for alternative production systems is a preferred method of presenting information on the economic problem to policymakers. The trade-offs between the various dimensions of sustainability are transparent and decision makers can place alternative weights on those dimensions in determining the appropriate balance between the health of the environment and the farm economy. Similarly, Pannell (1997) observed that simple approaches to sensitivity analysis, such as the trade-off curve approach, may actually be the absolute best method for the purpose of practical decision making.

An example of a trade-off curve is shown in Figure 6.1. The horizontal axis measures the level of environmental sustainability, which in this example is measured as greenhouse gas (GHG) emissions. The vertical axis represents net farm returns, while points A, B, C, and D correspond to four different farm management systems. System D is the most profitable system but it also generates the highest level of GHG emissions (GHG_D). In contrast, system A produces the lowest emissions (GHG_A) and lowest net farm returns (R_A). Systems B and C fall in between, but system C is inferior to B for it has the same level of GHG residuals (GHG_B) but is less profitable than B ($R_C < R_B$). Systems on the frontier of the trade-off curve represent the

Figure 6.1

Trade-offs between net farm returns and environmental sustainability indicator (greenhouse gas emissions) for four farm management systems (A, B, C, and D)

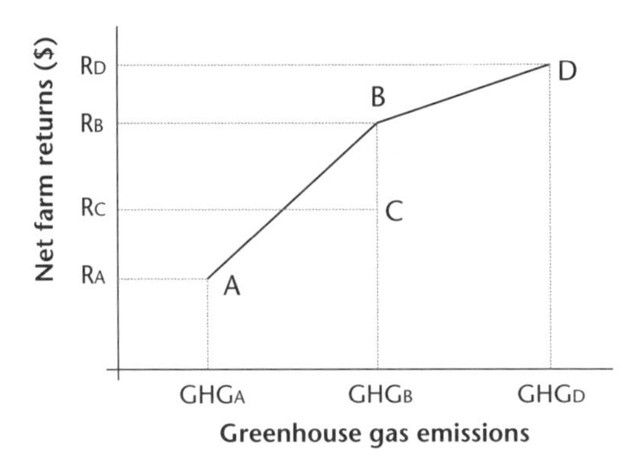

management choices that are "efficient" in producing a given level of environmental health for the greatest farm returns. Thus, the trade-off curve indicates which farming systems can cost-efficiently reduce GHG levels.

The trade-off curves also indicate the costs of achieving reductions in GHG levels at the farm level. For example, moving from system D to system B reduces GHG levels from GHG_D to GHG_B but also reduces farm returns from R_D to R_B. The difference in farm profits is referred to as abatement costs. A shift to management system A from B will reduce GHG levels by approximately the same amount as the initial shift from D, but the abatement costs are higher (R_B-R_A > R_D-R_B). The slope of the trade-off curve thus indicates the extent of the opportunity costs on one indicator when the other dimension of sustainability is changed (Yiridoe and Weersink 1998). Trade-off curves could also be developed to quantify the trade-offs associated with given levels of GHG emissions and other environmental sustainability indicators such as water quality, soil erosion, and biodiversity. This additional analysis would enable one to compare GHG emission reduction practices that are economically viable but perhaps impose some other environmental costs on the system.

In summary, trade-off curves provide a convenient means of summarizing information on the relationship between environmental and economic health for alternative agricultural production systems. The optimal level of the sustainability indicators is not selected as in the more complete economic framework; instead, outcomes are derived for alternative production choices. Policymakers can use this information to assess what mix of these competing sustainability dimensions is desired, and subsequently the associated set of management practices. Not only do the trade-off curves summarize the necessary information for making decisions on bigger issues related to agriculture but they also, as will be shown in the following sections, provide a means of conceptualizing the issues regarding sustainability and form the basis for empirical modelling of those issues.

The previous discussion assumed that the missing information in the economic problems related to agriculture and sustainability can be summarized conveniently through the use of trade-off curves. However, a fundamental step in defining the economic problem is to determine the dimensions of sustainability to be represented on the trade-off frontier. A precise operational definition of sustainability is lacking (Pezzey 1992; Zander and Kachele 1999) and is perhaps unattainable (Pannell and Schilizzi 1999). The choice of indicators of sustainability will therefore vary depending on the problem. The issue or issues related to sustainability for a specific problem arise through concerns expressed by affected stakeholders. The types of indicators related to agricultural sustainability and their linkages were discussed in Chapter 5.

A General Framework for Modelling Trade-offs

The framework developed in Chapter 5 provides a guide to designing and organizing research projects concerning the sustainability of agricultural production systems. However, the general framework that follows and the specific modelling questions discussed in the next section can also be adapted to other farm sector issues, such as the impacts of domestic support programs and international trade agreements.

The first step in the framework is to identify the issue, establish its context, and define the appropriate indicators for the issue, as outlined in the previous section. We find it useful in the initial assessment of the situation to consider the elements of a pollution problem as outlined by Tomasi and colleagues (1994). Various categories of information should be gathered, including whether it is a point source or diffuse source pollution problem, whether single or multiple firms are involved, and the objectives, choices, constraints, and knowledge of the polluting firms and the regulators. Particularly important, as will be seen later, is identifying the extent of spatial and technological heterogeneity. Incorporating spatial dimensions is necessary if damages vary by location with the same set of practices, while incorporating alternative firms in the model may be necessary if the profitability of firms, as well as their costs of abatement, vary across technologies.

After the current situation has been assessed and the issue(s) defined, the next step in the problem-solving framework is to identify potential goals or solutions to the problem. While the relationships between current management practices and the sustainability criteria were defined above, this step defines additional abatement technologies both at the extensive (types of inputs used) and intensive (rates of input use) margins. Acceptable levels for the sustainability criteria should also be ascertained in order to limit the scope of the modelling exercise. In addition, politically feasible policy options to address the sustainability issue should be identified. Initial analysis of a sustainability issue often focuses on the abatement costs of meeting alternative environmental objectives, with subsequent analysis addressing how the cost-effective technologies identified in the initial analysis can be promoted through regulations or economic instruments.

The final steps in the problem-solving framework are to formulate the models for estimating the levels of the sustainability indicators for the alternative technologies and/or policies identified previously. At the beginning of this phase, the scientists who can best provide the expertise to measure the sustainability criteria must be enlisted to support the project. Scientists' advice may be based on a quantitative model, on empirical data, or on their expert judgment. Where a model is used, the ideal is for it to be modified or developed with the overall purpose of the study in mind. Inevitably, however, the disciplinary models that are available have likely been predetermined by previous research efforts. This constraint exacerbates what is already a major difficulty: ensuring consistency of data between disciplines. Different disciplines tend to use different spatial and temporal units of analysis. For example, biophysical scientists tend to model smaller units (such as individual plants, fields, or animals) while economists tend to model larger aggregates, such as a farm or a geographic region. The dichotomy is not surprising given that disciplinary research tends to be guided by disciplinary orientation (reductionism), with little regard to how the information can be used to examine larger policy issues (Crissman et al. 1998).

There are two approaches for integrating different disciplines for the purposes of model development: interdisciplinary research and coordinated disciplinary research. Interdisciplinary research involves a high degree of interaction throughout the research process, as researchers from the relevant disciplines collaborate in planning and conducting a research project and analyzing the results. Coordinated disciplinary research involves researchers from relevant disciplines interacting primarily in the planning and analyzing phases of the research, while they work independently during the research phase itself.

Each approach has its advantages. Taking an interdisciplinary approach ensures consistency between the disciplines, and there is less risk of making incorrect and inappropriate assumptions. In contrast, taking a coordinated disciplinary approach enables individual researchers to make use of more advanced disciplinary tools and methods. Potential shortcomings of interdisciplinary research include the fact that it limits the scope of analysis for any one participating discipline and that it requires a greater degree of compromise between disciplines, making it more difficult to implement in practice.

The following section details the specific issues faced by economists in developing their farm-level model for sustainability issues. We argue that coordinated disciplinary research is the preferred approach in most cases, as advocated by Antle and colleagues (1998). Once the spatial and temporal units of analysis for all disciplines involved are resolved, collection of the necessary data can proceed. The final part of this evaluation step in the framework is the empirical integration of the different models in order to measure the values for the indicators under alternative management systems. Given that economists are trained to consider trade-offs, they have a comparative relative advantage in combining the research results to quantify the sustainability indicators and in presenting the results to policymakers. As argued in the previous section, trade-off curves highlighting the effects of alternative production practices on the appropriate sustainability indicators are an effective means of summarizing the results.

In summary, the problem-solving framework outlined in Chapter 4 is designed to ensure coordination of the disciplinary research necessary to address sustainability issues surrounding agricultural production systems. The first step ensures that there is agreement on the sustainability issue creating the need for research. Consensus on the problem and its importance then guides the researchers in identifying the appropriate sustainability criteria, measuring those indicators, and carrying out the integration process so that the trade-offs between those measures can be quantified. A major outcome is the consistent spatial and temporal unit of analysis between all scientists, which enables coordinated disciplinary research. Such data consistency is not often achieved, however. The choices and trade-offs facing economists when modelling farm-level decisions for bigger issues is discussed in the next section.

Trade-offs to Consider when Assessing Trade-offs

Once issues, indicators, and potential solutions are identified, the next step is evaluation of current conditions and assessment of alternative outcomes. This step requires the development of an appropriate modelling approach, which in turn requires the researcher to make choices in the formulation of the models. This section addresses the types of questions, issues, and decisions that need to be specifically considered by researchers interested in incorporating sustainability considerations into empirical farm-level modelling. The discussion is summarized in Figure 6.2.

Modelling Approach

In examining the relationships between the economics of agricultural production practices and sustainability, there are four categories of empirical

Figure 6.2

Questions to be considered by researchers in modelling farming system linkages

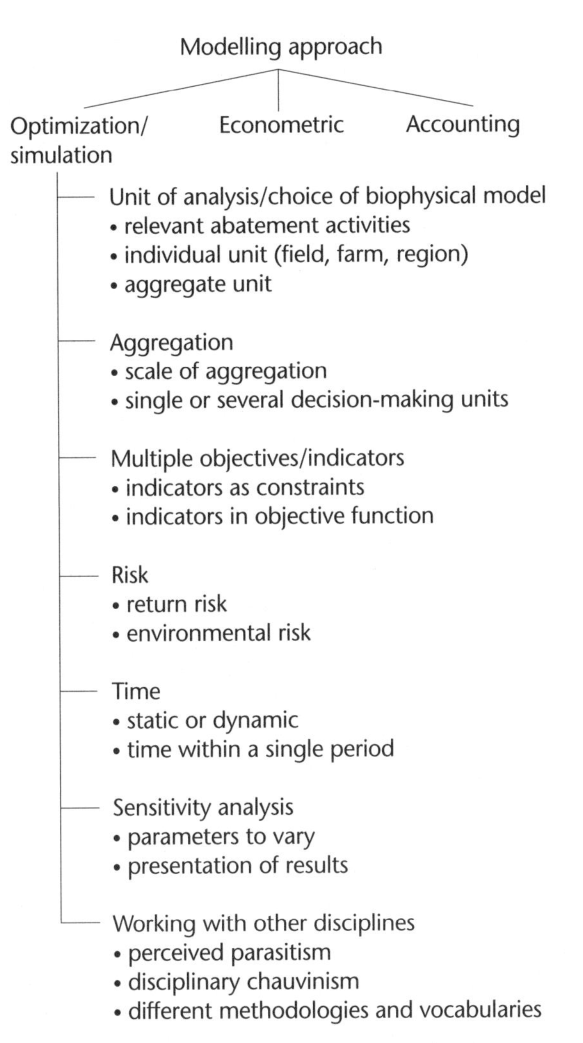

models that can be used: econometric, optimization, simulation, and accounting. Econometric models are statistical representations of farm-level systems, often estimated as aggregate systems of equations for input demand and output supply derived from duality theory. Optimization and simulation models are both systems of equations and/or inequalities designed to replicate farm-level activities related to production, marketing, finance, and so on. A distinction is often made between optimization and simulation models in that optimization models involve the specification of a behavioural assumption (such as profit maximization), while this is not the case for simulation models.[2] Finally, farm-level accounting models use budgets (capital, enterprise, or partial) to assess farm-level activities.

Econometric models enable statistical testing of economic and/or technical relationships. The ability to aggregate from individual units to a larger scale in a statistically consistent manner is another major advantage. Antle (1988) and Antle and Just (1991) first presented the framework for using econometric models to address the interaction of agriculture and its impact on environmental and human health. Subsequently, this approach has been used to examine sustainability issues with pesticide use in Philippine rice production (Pingali and Rogers 1995) and Andean potato production (Crissman et al. 1998). Farm management decisions such as pesticide use and land allocation at a small scale are modelled (i.e., estimated) as a function of prices, policies, technology, and physical characteristics of the site. A biophysical model is used to estimate environmental impacts on the site. The distribution of the physical characteristics for a region induces a joint distribution of the production choices and the environmental measures, thereby making it possible to aggregate those indicators to the regional level. The impact of altering prices and policies on the trade-off curve can be readily determined by simulating under alternative assumptions.

While this econometric approach can link farm-level decisions at the individual field level to aggregate sustainability indicators in a consistent manner, its disadvantage lies in its intensive data demands. Large amounts of information must be collected on the necessary variables such as input use, physical characteristics, and prices through a survey of individual farms in the region. Other, more aggregate econometric approaches use secondary data linking environmental health to input use (Wu and Segerson 1995), but these models cannot adequately capture many present and proposed farm-level decisions (such as application rate and timing of input use) that have an impact on sustainability indicators.

Optimization models have the advantage of providing the solution that best achieves the specified objective and, most importantly, enabling a detailed specification of farm-level activities. Simulation models also enable greater detail in specifying farm-level considerations, but with more flexible structures than are typically possible with optimization models. By their nature, however, these models are non-optimizing, and as a result there is no guarantee that the best option or solution is identified. Optimization

2 Farm-level simulation models also display similarities to accounting models, in that they typically involve the specification of accounting relationships such as profit, cash flow, and so on.

and simulation models permit one to analyze current technologies at the intensive and extensive margin but also other management practices.

The second step of the problem-solving framework in Figure 4.1 involves farm-level decisions about potential abatement technologies and their contributions to alleviating sustainability concerns. These choices will have to be included within the farm-level model in order for the sustainability issues surrounding agriculture to be assessed. Given the intensive data demands of the econometric approach, if it is to examine farm-level choices, the most common means of adequately examining the trade-offs are through mathematical programming and simulation models. Thus, the remainder of the discussion in this section focuses primarily on issues related to these types of models.[3]

Unit of Analysis/Choice of Biophysical Model

A major outcome of the problem-solving framework illustrated in Figure 4.1 is resolution of the decision about spatial unit of analysis. The size evaluated must be determined at the individual level where management decisions are made and at the aggregate level where the policy decisions are made.

The individual scale of analysis is based largely on the choice of the biophysical simulation model, which expresses changes in resource quality as a function of management actions and site characteristics. Reviews by Addiscott and Wagenet (1985) and Ellis and colleagues (1991) highlight the differences in focus, variation in scale of analysis, and data requirements for these models. Some are best suited for examining pesticide movement (LEACHP), while others, such as AGNPS or CENTURY, are geared towards assessing nutrient flow. In some cases, models that examine the same residual differ in terms of the management practices that can be considered. For example, in the case of examining means of reducing nitrate leaching into groundwater from agriculture, the CENTURY model was chosen by Yiridoe and colleagues (1997) over AGNPS since it can assess the effects of management choices such as fertilizer application rates on nitrate levels. The unit of analysis for CENTURY is an individual field. In another situation, the sustainability issue concerned the movement of phosphates via soil erosion into surface water. AGNPS was chosen since it allows for an area (such as a field, farm, or watershed) to be broken up into smaller units, with the movement of residuals from one unit to another affected by the management choices in the downstream unit. The effects of targeting programs and the value of the externality of the producers near the water body could

---

3 A good discussion of optimization models, including structure, assumptions, and applications, is provided by Hazell and Norton (1986) and Paris (1991). Law and Kelton (1982) provide a discussion of issues related to the use of simulation models.

be evaluated at the watershed level (Lintner and Weersink 1999). Provided the biophysical model can assess the practices suggested in step 2 of the problem-solving framework, the choice of model (and subsequently individual unit of analysis) will be determined by data availability and the model with which the natural scientists in the research team are most familiar.

The individual unit of analysis is dictated by the factors affecting the sustainability issues and the biophysical model chosen to represent the process. On the other hand, the aggregate unit of analysis depends on the political region affected by the issue. This larger unit may be as small as a farm if inferences can be drawn from an individual unit to a larger scale, or it may be at a local, watershed level as policy efforts to curb agricultural residuals are increasingly being initiated by municipalities. In most cases, however, legislative jurisdiction for the affected resource and subsequently the governing body will be at a state/provincial level, or at a national level in the case of a residual such as greenhouse gases.

Aggregation

As alluded to in the previous section, there are generally two units of analysis for consideration in farm-level modelling for bigger issues. A major decision facing researchers is how to combine or extrapolate the results examining the decisions of individual units to the larger scale for policy analysis. The greater the difference in sizes between the individual units and the aggregate, and the greater the heterogeneity within the individual units, the more complex the process of aggregating the results into the desired trade-off curve.

A method of aggregation with the mathematical programming approach is to assume that the individual units analyzed are regions and a social planner is maximizing returns for all regions subject to environmental quality constraints. If the decisions made at the regional level have an impact on prices, then the farm-level model(s) may need to be combined with a broader, sectoral model that endogenizes price. An example of such an approach is the work by Kulshreshtha and colleagues (1999), who use a model that maximizes national producer and consumer welfare through the choice of regional crop and livestock activities, and combine it with a simulation model that predicts greenhouse gases from the selected activities.

Another approach that can be used is to select and model activities for a set of representative farms. The results for these farms are then "aggregated up" based on information about actual distribution of farms within a region. For example, the representative farms could be categorized by size, by subregions, or by different mixes of available resources. The results for the farms could be aggregated based on the actual distribution of the farm categories in the region.

Multiple Objectives/Indicators

The development of a mathematical programming model requires that an explicit objective function be defined reflecting the decision maker's behaviour or goals. In a simulation model, there is no explicit objective function. However, the results of simulation models are often evaluated on the basis of the assumed objectives. The behavioural assumption of profit maximization or cost minimization typically relates to one of the sustainability indicators, and the other indicator on environmental health is incorporated within the model as a constraint and/or activity. Altering the exogenous variables related to prices, policy, or technology enables one to determine the trade-offs between the indicators.

It has been assumed that the trade-off between the multiple indicators of sustainability is sufficient information for policymakers, who subjectively evaluate the opportunity costs of improving environmental health. However, an alternative to specifying one indicator within the objective function and the others as constraints is to include all indicators within the objective function. One approach that may be used to deal with multiple objectives is goal programming (GP). GP models allow more than one objective to be optimized simultaneously through the specification of targets for each goal, with the mathematical objective then being to minimize deviations from those targets (Romero and Rehman 1989). Incorporating different weights for each goal, allowing prioritization of the multiple objectives, can enhance the approach. GP modelling adds to the size and complexity of the farm model and requires the specification of targets (and possibly weights) for each goal. Nonetheless, it does provide a tool that enables farm management researchers to incorporate very different decision-maker goals within a single model.

In the case where sustainability indicators can be identified but are difficult to quantify, an alternative modelling approach of potential value is nearly optimal linear programming (NOLP) (Jeffrey et al. 1992). As the name might suggest, NOLP is an applied modelling approach producing solutions that are not optimal with respect to any one objective but instead are somewhat or "nearly" optimal for all objectives. There are different algorithms by which NOLP may be empirically implemented. In general, they involve solving an initial farm model, using a relevant economic goal (such as profit maximization) as the objective function, to obtain an "optimal" solution. A search procedure, which varies by NOLP algorithm, is then used to identify alternative solutions that are nearly optimal in the sense that they are within some specified tolerance of the original optimal solution.

The intent of NOLP is to search for alternative solutions that are almost optimal from the perspective of the economic objective, but that may be "better" than the original solution in terms of other objectives, which might be, for example, non-quantifiable sustainability indicators. The advantage of NOLP is that it does not require the researcher to specify targets or weights for the alternative objectives. Instead of a single solution, NOLP typically results in several possible solutions that may be presented to decision makers. Essentially, trade-offs between goals are evaluated by the decision makers outside of the farm model, as opposed to GP, where targets and weights are used to evaluate trade-offs between goals within the farm model.

Both GP and NOLP have been used to address multiple objectives in empirical farm-level analysis (e.g., Rehman and Romero 1984; Minguez et al. 1988; Jeffrey and Faminow 1995). As the emphasis on incorporating environmental or sustainability considerations in farm-level models increases, it is likely that these types of modelling approaches will be used more commonly by applied researchers working in this area.

Time

Farm-level analysis is typically conducted using a static model; that is, an annual model of production is developed where the results reflect the activities of a representative farm operation for a representative year. This practice is often carried over to applications involving sustainability issues. However, there are instances when the modeller requires a more dynamic representation of farm-level decision making, with time explicitly incorporated into the analysis.

There are two relevant aspects of time when considered in the context of farm management research. Time may be considered in terms of multiple decision-making periods, most often represented by extending the analysis over more than one year.[4] Additionally, time may be relevant within a single decision-making period; that is, the year may be divided into multiple time periods (e.g., months).

Both aspects of time (multiple years and multiple periods within a year) become potentially more important when considering environmental and/or sustainability issues. For example, it is difficult to accurately assess the long-term impacts of current farm production practices on environmental variables (e.g., groundwater quality) within a purely static model. Just as important in some cases may be the implications of timing within the year for certain agricultural production practices, in terms of the impact on environmental variables. For example, the timing of pesticide or fertilizer application, or the disposal of manure for a livestock operation, may affect the degree to which these practices impact on the environment.

4 Hazell and Norton (1986, ch. 4) and Rae (1994, ch. 7) discuss the implications of including this aspect of time in optimization models.

While it might then seem apparent that incorporating dynamics is necessary, this issue actually represents another instance of evaluating trade-offs. The "cost" of incorporating dynamics is seen in increased model complexity through the need for additional activities, constraints, and dynamic linkages that connect the different time periods in the model. In contrast, less complex static models are often still able to provide useful contributions in the analysis of sustainability issues, as indicators of producer behaviour. As well, in some cases dynamics may be approximated within a static model through the appropriate definition of activities or constraints. For example, this could include defining cropping activities as crop rotations rather than individual crops – that is, acres of wheat-wheat-canola rotation rather than acres of wheat (e.g., Kingwell and Pannell 1987).

Risk

Risk is pervasive in agriculture and natural resource management, and most farmers are averse to risk (e.g., Antle 1987; Binswanger 1980; Bond and Wonder 1980). However, representing risk adds to model complexity and it may or may not be sufficiently important to be worth modelling risky variables and risk attitudes explicitly. Pannell and colleagues (2000), in the context of farm modelling, argue that it is often not worthwhile, for the following reasons:

- It often makes less difference to identification of the optimal strategies than correct representation of underlying technical relationships, which are therefore a higher priority for use of scarce modeller time.
- If the purpose of the model is normative (recommendation of optimal strategies) rather than positive (prediction of farmer behaviour), the difference in value of recommended strategies from models that do and do not represent risk aversion is, in most cases, extremely small.

These conclusions probably apply also to sustainability-related issues. Therefore, for practical analyses, it may often be sufficient to employ sensitivity analysis (Pannell 1997) to explore consequences of alternative risky outcomes, rather than modelling the relevant probability distributions endogenously.

Of course, it is possible to envisage situations where explicit modelling of risk and uncertainty would be desirable. For example, suppose a researcher wishes to assess the influence of public safety net programs (e.g., crop insurance) on "sustainability," however sustainability is to be measured. In this case, given the assumptions underlying the existence of the public programs, incorporating technical and market risk may be considered necessary to accurately predict farmer behaviour. Even here, however, we would expect direct financial incentives to outweigh risk considerations in shaping farmer responses.

If environmental questions and problems are being modelled, perhaps the most relevant source of uncertainty to be modelled is "environmental" risk. This may include such risks as the possibility of a weather event causing failure in a manure storage system or significant runoff of pesticides into a water system. In these instances, the risk may be best quantified not as a distribution characterized by a mean and variance but as a probability of "disaster."

Assuming that the modeller does decide to represent risk and risk attitudes, there are a number of techniques available. The most common methods of assessing the trade-off between risk and return have been empirical models such as quadratic programming, MOTAD, and Target MOTAD. These models are easily able to incorporate empirical risk distributions or distribution parameters in order to evaluate the trade-off between technical and/or market risk, and return (often measured as profit). In their review of available risk programming methods, however, Hardaker and colleagues (1991) conclude that the DEMP (Lambert and McCarl 1986) and UEP (Patten et al. 1988) approaches have strengths that make them preferable. They are special cases of discrete stochastic programming, which has the advantage of being able to represent within-year tactical decision making, and of explicitly representing probabilities of adverse environmental outcomes. Less flexible, but also enabling the analyst to optimize some economic objective (e.g., maximize profit) subject to a maximum probability of environmental damage, is the method of chance constrained programming (Charnes and Cooper 1959). Its simpler approach may be preferred in some cases.

Simulation is also a valuable tool for studying risk. Monte Carlo simulation methods, which are now widely available as add-ons for spreadsheet programs, are ideal for generating probability distributions needed by the programming methods described above. In addition, they are valuable analytical approaches in their own right and are increasingly being used.

Conclusions

The cause-and-effect relationship between agricultural production systems and environmental health must be incorporated into any analysis designed to provide input on sustainability issues surrounding agriculture. This chapter has provided suggestions for researchers attempting to assess aspects of the "sustainability" of alternative agricultural production systems. Trade-off curves represent a convenient means of summarizing the information for policymakers and form the basis for conceptualizing and empirically modelling issues regarding sustainability.

A major issue facing modellers of agricultural sustainability is the difficulty in ensuring consistency of data between disciplines. The unit of analysis both at the individual decision-maker level and at the aggregate policy level should be defined on the basis of the important economic issues regarding sustainability. Any biophysical model used to estimate the effect of agricultural practices on resource quality must be able to account for the intensive and extensive management choices that are felt to be contributing to the problem of concern. The individual unit of analysis tends to be determined by the unit used in the best available (i.e., the most appropriate) biophysical model. The researcher then must construct a decision-making model for this individual unit, and in the process consider many of the issues that are common to farm-level models designed to enhance individual returns, such as time and risk. However, the modeller must also consider how to aggregate the results across heterogeneous units to the level at which policy decisions are made, and how to handle multiple objectives.

Part 3: Applications of the Framework and Linkages for Solving Sustainability-Related Problems of Intensive Agriculture

7

A Whole-Farm Systems Approach to Modelling Sustainable Manure Management on Intensive Swine-Finishing Farms

D.P. Stonehouse, G.W. de Vos, and A. Weersink

As alluded to in Chapter 1, a multiplicity of factors has been behind the agricultural intensification trends of recent decades.[1] Globalization and increasing competition from international trade agreements have forced farmers to be more cost-conscious. Technological progress and capital-labour substitution have provided the means to become more cost-competitive, and also more specialized, with increasing emphasis on larger operations. One result has been the emergence of intensive livestock operations (ILOs), noted for their size, their specialization in a single type of livestock, the density of livestock in relation to the land base, and, in some cases, problems with manure disposal. Economic aspects of sustainability (profitability and international competitiveness) have apparently superseded environmental aspects of sustainability (managing the farm business to ensure minimum pollution damage, especially from manure operations).

Controversy surrounds the handling of livestock manure on North American farms, but most especially in ILOs. Growing public concerns about environmental protection needs are obliging farmers to become increasingly careful of how manure is handled in order to obviate, or at least minimize, leakage of potentially harmful plant nutrients and microorganisms. Hog operations in southern Ontario, where there is growing urbanization and industrialization, are under particularly close scrutiny. Managed appropriately, livestock manure supplies an important part of the farmer's crop nutrient requirements without damaging the environment. Farm decision makers must therefore balance the technical (biophysical) aspect of manure

1 The generous financial support for this research from the Ontario Ministry of Agriculture, Food and Rural Affairs, from the Canadian Farm Business Management Council, and from Ontario Pork is gratefully acknowledged. This chapter is patterned on a previous journal article, but with substantial modifications: D.P. Stonehouse, G.W. de Vos, and A. Weersink, "Livestock manure systems for swine finishing enterprises," *Agricultural Systems* 273 (2002): 279-96, with permission from Elsevier Science.

as a source of plant nutrients with the environmental aspect of avoiding damage. At the same time, management practices designed to maximize plant nutrient supply from manure and/or contain environmental impairment tend to be costly. Also, the benefits of such practices do not always offset the costs to the farmer, so that decisions take on a three-way trade-off perspective, involving the biophysical, environmental, and economic aspects of sustainability.

Moreover, manure management in a practical context must be thought of as only one facet of overall farm management. Manure-handling operations must blend with livestock and crop production and marketing plans, rather than being separate entities. While previous research efforts have incorporated the three-way trade-offs among biophysical, environmental, and economic aspects of manure handling, little has been done to link manure management with overall farm management. This chapter integrates manure management with crop production and livestock management in a whole-farm setting, thereby showing how manure management decisions and crop and livestock production decisions influence and support each another.

The objective of this study was to develop a decision-making aid for assessing the biophysical, environmental, and economic merits of alternative manure-handling systems in the context of a whole-farm planning model. Specifically, a computerized decision tool for specialized swine-finishing ILOs in Ontario was envisaged. Ontario's rapid pace of economic development has led to growing collisions between urbanization, industrialization needs, and agricultural intensification needs on a shrinking farmland base. With an expanding swine industry and a restructuring towards ILOs, swine operations have become targeted in southern Ontario as potential environmental polluters. Hog farmers were seen as major beneficiaries of an analytical tool designed to evaluate the biophysical, environmental, and economic impacts of alternative manure systems for intensive swine-finishing operations of varying sizes in ever-closer proximity to urban industrial centres. Trade-offs among biophysical, environmental, and economic aspects of sustainability for swine-finishing ILOs were envisaged as a principal outcome, leading to alternative solutions or courses of action, as depicted in Figure 4.1.

Research Methods

Simulation versus Optimization Models

Previous studies have largely relied on either simulation methods or optimization techniques to evaluate the relative merits of manure systems alternatives. Simulation methods employ systems of equations to model real-world relationships under a range of different biophysical, environmental, or economic circumstances. Biophysical examples have included manure production and composition from different livestock types (American Society of Agricultural Engineers 1998; Tuitock et al. 1993), the fate of plant nutrients under alternative manure-handling systems (Sommer et al. 1995; Gordon et al. 1988), and rates of nutrient availability and uptake by different crops (Barnett 1994; Paul and Beauchamp 1993). Environmental examples have investigated gaseous emissions from manure (O'Halloran 1993; Paul et al. 1993) and leaching into water bodies of manurial N (Martins and Dewes 1992), manurial P (Sharpley and Halvorson 1994), or manurial bacteria (King et al. 1994). Simulation examples from economics have typically employed cost-benefit analysis for comparing alternative manure-handling systems (Fulhage 1994) or for estimating costs to farmers of adopting environmentally sound manure practices (Baltussen and Hoste 1993). Some simulation studies have combined the biophysical, environmental, and economic aspects of manure management into a single decision support system model (Ogilvie et al. 2000; Chapter 8). Whether focused or comprehensive in scope, simulation approaches are useful for depicting actual outcomes of manure management decisions, and can therefore provide a priori guidance to farm decision makers.

In contrast, optimization techniques, based on mathematical programming (MP) models, seek the best possible outcome to a broad range of biophysical, environmental, and economic circumstances. They can point the way to improvements in existing situations by determining the minimum-cost manure management system for the different types and sizes of livestock enterprises (Boland et al. 1999; Kelland and Stonehouse 1984). MP models can also be used to determine how manure systems best fit into overall farm crop-livestock systems (Roka and Hoag 1996), and how best to fashion policies to encourage environmentally safe manure management (Fleming et al. 1998; McSweeny and Shortle 1989). All such optimal solutions are obtained subject to resource availabilities, farmer preferences, government policy dictates, or other constraints. An added bonus from MP model solutions stems from the identification of operative constraint(s), the generation of shadow values (marginal value products) attached to resources in scarce supply, and the range of farm activity levels over which the optimal solution remains stable.

For this study, optimization techniques were selected because their search for the most economically sustainable outcome under environmentally constrained circumstances was thought to parallel farmers' decision-making goals and procedures. Second, these techniques lend themselves well to the analysis of biophysical, environmental, and economic trade-offs in a sustainability context. That is, resources must be used efficiently for profit maximization to be realized, but excess nutrient application to either plants or animals can not only lead to waste but also damage the environment.

Third, MP models are well adapted to the needs of environmental policy evaluations through comparisons between constrained and unconstrained pollutant emissions outcomes.

Conceptual Framework

Modelling manure-handling alternatives integral to a whole-farm planning situation sufficient to consider biophysical, economic, and environmental sustainability factors requires an interdisciplinary approach. At stake are decisions about livestock production systems and feed alternatives affecting animal performance, and manure production and composition; manure collection, storage, field transfer, and application alternatives; crop production systems and plant nutrient management alternatives affecting nutrient uptake and crop performance; environmental damage potential from manure-handling alternatives; and business performance in the form of resource use options and costs, revenues generated, and profits realized. These parameters and their interrelationships can be depicted schematically (Figure 7.1). Such a complex set of factors for the present study necessitated close cooperation among members of a research team consisting of animal nutritionists, land resource scientists, plant nutritionists, hydrological and systems engineers, and agricultural economists.

Livestock Production

The focus in this study was on intensive hog-finishing enterprises. Growing-finishing pigs typically account for 75% of feed usage and manure production in integrated swine farrow-to-finish farms; therefore, nitrogen (N), phosphorus (P), and potassium (K) balances for growing-finishing pigs closely resemble those for the integrated swine farm (de Lange and Porteaux 1999). Weaner pigs are purchased at an average liveweight of 25 kg and marketed as finishers 100 days later at an average weight of 108 kg.

Four enterprise sizes were modelled, based on swine accommodation capacities of 200 (small), 500 (medium), 1,000 (large), and 5,000 (extra large) per 100-day cycle. Three cycles per year were assumed. Sales of finished pigs constituted one of only two sources of revenue, the other being sales of crops. Costs were incurred for weaner purchases, feeds, labour, and amortization of livestock housing capital outlays.

Manure Production and Composition

To simplify the calculation of increasing manure output as the hog grows from 25 kg to 108 kg, it was assumed that each hog averaged 0.0045 m^3 per day, or 0.45 m^3 per 100-day cycle. Total manure produced per annum is then the product of daily manure output per pig, number of pigs per enterprise, and number of cycles per year.

Figure 7.1

Schematic diagram of management decisions affecting economic (costs and revenues) and environmental (GHG emissions, excess P, excess N) sustainability factors on intensive swine-finishing farms

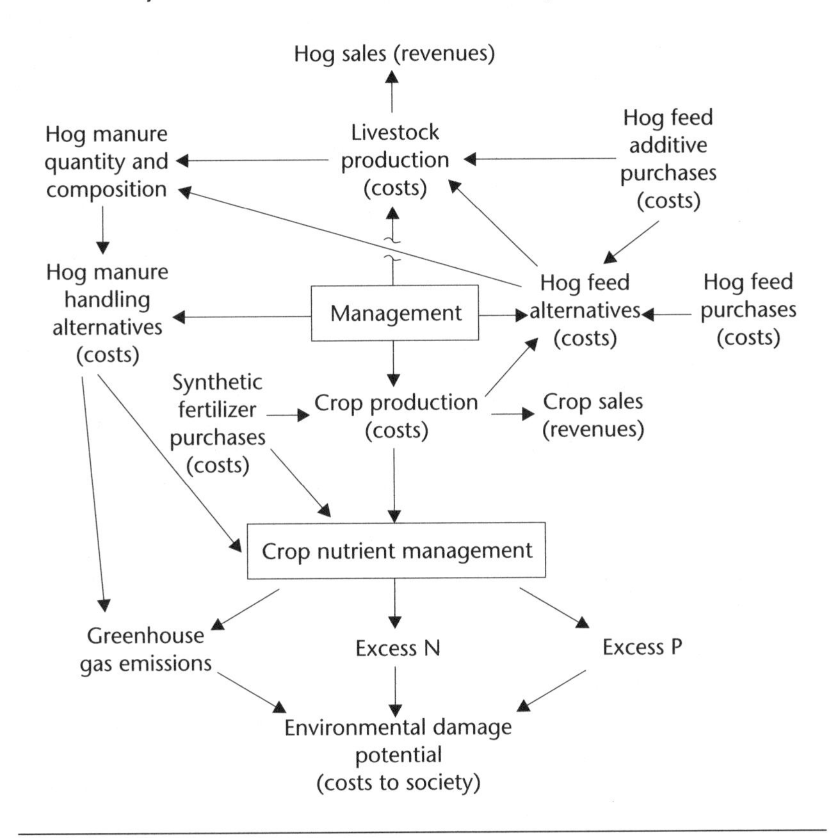

Composition of manure is a function of feed ration constituents and their nutrient content, feed nutrient retention rates by the pig, feed wastage, and average carcass lean yields, as well as starting and finishing weights of the pig. For this model, calculation rules were developed for predicting excretion rates of N, P, and K for different classes of swine (de Lange 1999).

Manure Handling Alternatives

Manure handling can be suitably categorized into collection, storage, and field application components. For our modelling work, it was assumed that all manure would be handled in liquid form, in accordance with recent Ontario trends (Ho et al. 2000), and that each of the three components would have three handling alternatives as follows:

- collection by gravity through fully slatted floors to below-floor temporary storage; by gravity through partially slatted floors to below-floor temporary storage; by manual scraping of solid concrete floors into gutter cleaners, thence by pump to storage
- storage, with a capacity sufficient for holding manure from twelve months' production, in a rectangular earthen pit; in a rectangular, above-ground, open-topped concrete tank; in a rectangular, above-ground, covered concrete tank
- field application either on owned land or on land rented from a neighbour by self-reel-in travelling irrigation gun with subsequent soil incorporation within five days of application; by tanker broadcaster with subsequent soil incorporation within twenty-four hours of application; by tanker injector.

Any combination of each of the three alternatives for each of the collection, storage, and field application components is possible, making for twenty-seven alternative ways of handling manure. Each of the nine alternative handling components carries with it predetermined unit operating costs (for fuel, machinery repairs, labour); ownership costs (amortized capital outlays for machinery, equipment, and building components); and ammonia (NH_3), nitrogen (N), and phosphorus (P) loss rates (Manure Systems Research Group 1999). For each of the twenty-seven manure-handling alternatives, therefore, operating and fixed costs and NH_3, N, and P losses become additive across handling components. Operating costs are a function of hog enterprise size, which in turn dictates equipment or building facility size, as well as manure volume. Fixed costs are a function of enterprise size only, which in turn determines equipment or building facility size and therefore capital outlay. NH_3, N, and P loss rates are dependent upon manure volume and choice of manure-handling system.

Crop Nutrients Available

The macro plant nutrients N, P, and K, expressed in equivalent fertilizer nutrient forms as N, P_2O_5 (phosphate), and K_2O (potash), respectively, were modelled essentially as nutrient management plans (NMP), so that nutrient supplies (from all sources), nutrient requirements by the growing crop, and all types, sources, and extents of nutrient losses are recorded. Sources of nutrients are soil inventories, livestock manure net of all losses during handling, and supplementary synthetic fertilizer. Demands for nutrients are a function of crop grown and expected yield. Allowance is made for manurial N being available 60% as ammoniacal N (100% immediately available to the plant) and 40% as organic N, with a 30% mineralization rate for plant availability in the year of manure application (Manure Systems Research Group 1999). Manurial P is converted to P_2O_5 by multiplying by 2.29, and then by 0.4 to reflect 40% availability in the year of manure application. Manurial K is multiplied by 1.2 for conversion to K_2O, and then by 0.9 to reflect 90% availability in the year of application.

Manure application rates, assumed to be exogenously set by the farm operation in accordance with good NMP practice, may nevertheless lead to excess or deficit in one or more macro nutrients: one can balance for N or P or K needs, but not all three simultaneously. Where nutrient deficits occur, supplementary synthetic fertilizer can be purchased in the model at market cost. Where excess nutrient is indicated, supplementary cropland can be rented in at market cost to absorb the manure and therefore the nutrient excess. Otherwise, N and P nutrient losses into ground and/or surface water bodies are calculated in the model and are assumed to contribute to environmental pollution damage (at a cost to society). Surplus K may also be leached into water bodies, but because this is not considered potentially environmentally damaging, it was not included as an environmental sustainability indicator in the model. The option to rent supplementary cropland for manure disposal purposes would incur, in addition to land rental costs, increased manure transportation charges for greater distances shipped.

Crop Production

A common problem on Ontario intensive hog farms is shortage of land for manure disposal and feed crop production purposes. This was taken into account in our study by assuming that, for each of the four hog enterprise sizes, owned tilled land could supply an average of 62% of the hog feed corn requirements. The resulting tillable land base was therefore limited to a maximum of 8, 20, 40, and 200 ha for hog enterprises marketing 600, 1,500, 3,000, and 15,000 finishers per annum, respectively. Tillable land can be used to grow corn or soybeans, each of which can be sold at market prices or retained for feeding the hogs.

The same tillable land received all liquid manure from the hogs, constrained to be applied in the spring between snowmelt and field operations for seedbed preparations and crop planting. With excess plant nutrients from manure a distinct prospect, presumably most manure would be directed towards land for producing corn, with its greater nutrient absorption capacity than that of soybeans.

Feed Alternatives

Ontario hog-finishing operations characteristically feed a corn-soybean meal ration, based on the two commonest field crops. While phase feeding by growth stage of the hogs is popular in practice, a single-phase feeding strategy was assumed here for reasons of simplicity. Two variants of the corn-soybean meal ration were offered as alternatives. One variant includes synthetic lysine, one of the essential amino acids in the pig's diet. The added lysine can replace some of the crude protein in the ration and hence reduce nitrogen emissions in hog manure. The added lysine can also help reduce phosphorus emissions in the manure.

The other variant conjoins exogenous phytase with the corn-soybean meal ration. The added phytase can reduce phosphorus requirements, and hence phosphorus emissions in hog manure. A feed ration containing either additive could be important in meeting manurial N or P constraints under an NMP protocol. Moreover, each additive has been shown to increase feed efficiency in hogs (de Lange 1999).

As previously discussed, corn and soymeal feed ingredients can be obtained from farm-produced crops, but not totally. Supplemental purchases are necessary at market prices. Alternatively, all farm-produced crops can be sold at market prices and all hog feed ingredients can be purchased at market prices.

Management Decisions

The principal decisions for managing a hog-finishing enterprise relate to the following (see Figure 7.1):

- hog enterprise size, including accommodation capacity per cycle and per annum, and the extent to which capacity is used
- choice of manure-handling components for collection, storage, and field application, including storage capacity and timing of field applications
- crop nutrient provision from manure and/or synthetic fertilizers, keeping in mind the extent of land base, owned and rented; the capacity of the land to absorb nutrient supply; and the fate of any excess nutrients according to NMP requirements
- allocation of tillable land to alternative crops, and the usage of those crops as feeds for livestock or sale
- hog-feeding strategy, including choice of food ingredients, keeping in mind impacts of ration fed on manure volume and composition, and own-farm versus imported sources of supply.

The successful decision agent presumably recognizes that all these decisions have impacts on each other, that all play a role in an integrated systems approach to hog production, and that all contribute to the overall goal achievement of the business. While the goal was assumed here to be maximization of farm profits, a goal of environmental protection could just as easily have been assumed, using any one or all three of the NH_3 emissions, excess N, and excess P sustainability indicators. The focus in this study was on predictive indicators only, rather than summative and predictive, because no attempt was made to link on-farm manure activities with off-farm environmental damage. Sustainability indicators were therefore farm-family income (predictive for economics), soil nitrogen flux, building ammonia losses, and surplus phosphorus load (predictive for environment), as per Figure 5.2 in this book.

Empirical Model

As previously stated, optimization techniques were argued to be appropriate analytical tools for this study. Linear programming (LP) models lend themselves well to the task of analyzing multiple decision options in an integrated whole-farm planning setting such as that contemplated here. A tendency of LP models is to select mixtures of systems components in pursuit of their bid to maximize objective function values, in this case, maximization of farm net returns (profits). Such mixtures of components are quite acceptable in some instances, as, for example, in the case of both owning and renting land, or both purchasing and producing one's own livestock feed ingredients. In other instances, component mixtures would not be acceptable, as in the case of manure-handling systems or livestock feed rations. Only one method of collecting manure would be employed typically on any one farm, and that single collection method would be coupled with one storage and one field application method. It would be unusual to find on any one farm both irrigation and tanker-injector methods for field application of manure. Similarly for feed alternatives, one farm would likely use either standard corn-soybean meal or corn-soybean meal with synthetic lysine or exogenous phytase added, but not all three.

To avoid the inclusion of unwanted multiple systems components, mixed integer programming (MIP) models were used in this study. An extension of LP models, MIP models set selected activities (columns in the models) as integer variables. Integer variables have activity levels bounded at the upper level of 1 and the lower level of 0, so that such variables are either included fully in the basic feasible solution or excluded entirely (Ignizio and Cavalier 1994).

Stated more formally, the MIP problem is defined (according to Hadley 1964, with adaptations) as follows. Find a non-negative vector of real (manure handling, crop production and nutrition, animal production and feeding) activities and integer activities for manure systems components and hog feed ration choices,

$$x \geq 0,$$

that satisfies a set of m simultaneous linear equations reflecting resource availabilities, environmental damage, or other constraints,

$$\leq \geq b,$$

and that satisfies the constraint set associated with the subset, x_j, of integer activities,

$$j \varepsilon J,$$
$$x_j - d\, \delta_j \leq 0,\ j = 1, \ldots, n$$
$$0 \leq d_j \leq 1,\ j = 1, \ldots, n,\ \delta_j \text{ integers,}$$

and that maximizes overall farm net returns, inclusive of environmental damage containment costs,

$$(A_j \delta_j + c_j x_j)$$

By declaring all nine manure systems alternative components (three collection, three store, three field applications) and all three livestock feed alternatives as integer variables, we ensured that only one of the eighty-one possible manure-feed combinations would be selected at a time for any one run of the MIP model. In a practical vein, this reflects the need to pay annual amortization costs on only one component each for manure collection, storage, and field application, and only one feeding system. For analytical purposes, such a model structure enabled us to compare each of the eighty-one manure-feed combinations and to rank them, economically and environmentally.

The remaining variables, specified as real (number) activities in the model, catered to management decisions required for hog production levels, and associated manure production, synthetic fertilizer purchases, cropland rentals, crop production and sales, and feed purchases.

Row constraints were specified for limiting hog enterprise size; handling all hog manure produced; accounting fully for all N, P, and K macronutrients in the manure from initial supply through handling losses and soil mineralization to plant availabilities; limits to own cropland and rented cropland available; disposition of crops to feed or sales; meeting N and P macronutrient application standards limits; meeting maximum total manure application limits; accounting for excess N, P, and K applications (and, implicitly, losses to the environment).

Data Sources

A set of four hypothetical hog-finishing farms, one for each size category, was modelled as opposed to actual case situations. Consequently, secondary data sources were relied upon for all market prices for purchase, sale, and rental activities; crop production costs; crop yields; and hog feed conversion ratios (Ontario Ministry of Agriculture, Food and Rural Affairs).

Secondary sources also supplied data for manure production rates; manure composition; retention/loss rates for NH_3, N, and P for alternative manure collection, storage, and field application components; soil mineralization rates for N and P; crop nutrient absorption rates; manure-handling components operating costs and annual amortization of capital costs (Ogilvie et al. 2000; Manure Systems Research Group 1999; Goss et al. 1996).

Results and Discussion

For each of four hog-finishing enterprise sizes (200, 500, 1,000, and 5,000 hogs per 100-day cycle), a total of eighty-one runs of the MIP model was performed, one for each possible feed/manure-handling combination. Here we have presented results for only four runs for each enterprise size category, one for each of the four predictive indicators for sustainability. That is, there is one run to show the top-ranking economic whole-farm plan, and one run each for the top-ranking environmental whole-farm plan based on minimum NH_3 losses, minimum excess N, and minimum excess P. This allowed emphasis to be placed on the trade-offs among all four sustainability indicators, and hence the need for compromise in decision making, in the manner explained in Chapters 4 and 5.

Certain elements of the optimal whole-farm plans were common across all eighty-one model runs for each size category:

- Each enterprise size category operated at full hog capacity.
- All available owned cropland was allocated to corn production.
- Hog manure was disposed of exclusively on own corn land, and no neighbouring croplands were rented for manure disposal purposes.
- All farm-produced corn was sold at market prices.
- All feed ingredients (corn, soybean meal, mineral/vitamin supplement, and feed additives where applicable) were purchased as separate components at market prices.

Beyond the common elements, optimal solutions displayed a good deal of diversity in choice of feed/manure-handling components across hog enterprise size categories and economic versus environmental criteria. Four optimal solutions were generated for each of the four farm sizes (Tables 7.1 and 7.2). The optimal solutions are given for maximizing farm net returns (FNR) (in column 9, Tables 7.1 and 7.2), and for minimizing environmental impacts in terms of ammonia, excess N, and excess P separately. Under the economic criterion of maximization of farm net returns, feed ration 2 was consistently selected across all size categories (see first row results under each size category, Tables 7.1 and 7.2). This is the corn-soymeal ration with added synthetic lysine, which decreases N and P emissions in the manure and also improves hog feed conversion ratios. The economic criterion called for a manure-handling system based on solid floor collection, earthen pit storage, and travelling irrigation gun for field application across small,

Table 7.1

Optimal economic versus optimal environmental whole-farm plans for small and medium-sized hog-finishing enterprises in Ontario

	Manure management system								Environmental ranking		
Feed ration	Collect	Store	Apply	Farm net returns (FNR) ($)	Ammonia loss (kg)	Excess N (kg)	Excess P (kg)	Economic ranking (FNR)	Ammonia	Excess N	Excess P
SMALL: 200 hogs/100-day cycle; 600 hogs/yr; 8 ha tillable land											
2	SF	EP	I	24,694	1,029	212	205	1	74	4	2
2	FS	CC	TI	22,383	467	503	205	42	1	62	2
2	SF	CC	I	23,025	909	141	205	24	58	1	2
3	SF	EP	I	23,907	1,124	321	95	13	80	17	1
MEDIUM: 500 hogs/100-day cycle; 1,500 hogs/yr; 20 ha tillable land											
2	SF	EP	I	61,819	2,572	530	511	1	74	4	2
2	FS	CC	TI	57,020	1,167	1,257	511	42	1	62	2
2	SF	CC	I	58,639	2,272	352	511	23	58	1	2
3	SF	EP	I	59,852	2,811	803	237	15	80	17	1

Feed ration: 1 – standard corn and soybean meal; 2 – standard ration with supplemental synthetic lysine; 3 – standard ration with supplemental exogenous phytase.
Collect: SF – solid floor with gutter cleaner; FS – fully slatted floor with gravity collection.
Store: EP – rectangular earthen pit; CC – rectangular, above-ground covered concrete tank.
Apply: I – self reel-in travelling irrigation gun, soil incorporation within 5 days of application; TI – tanker injector.

Table 7.2

Optimal economic versus optimal environmental whole-farm plans for large and extra-large hog-finishing enterprises in Ontario

Manure management system									Environmental ranking		
Feed ration	Collect	Store	Apply	Farm net returns (FNR) ($)	Ammonia loss (kg)	Excess N (kg)	Excess P (kg)	Economic ranking (FNR)	Ammonia	Excess N	Excess P
LARGE: 1,000 hogs/100-day cycle; 3,000 hogs/yr; 40 ha tillable land											
2	SF	EP	I	123,600	5,143	1,060	1,023	1	74	4	2
2	FS	CC	TI	115,620	2,335	2,514	1,023	41	1	62	2
2	SF	CC	I	118,600	4,543	703	1,023	21	58	1	2
3	SF	EP	I	119,663	5,622	1,606	473	15	80	17	1
EXTRA-LARGE: 5,000 hogs/100-day cycle; 15,000 hogs/yr; 200 ha tillable land											
2	PS	EP	I	615,069	23,457	6,659	5,114	1	64	9	2
2	FS	CC	TI	585,733	11,675	12,572	5,114	38	1	62	2
2	SF	CC	I	590,855	22,717	3,516	5,114	30	58	1	2
3	PS	EP	I	595,403	25,639	9,511	2,366	20	73	30	1

Feed ration: 1 – standard corn and soybean meal; 2 – standard ration with supplemental synthetic lysine; 3 – standard ration with supplemental exogenous phytase.
Collect: SF – solid floor with gutter cleaner; PS – partially slatted floor with gravity collection; FS – fully slatted floor with gravity collection.
Store: EP – rectangular earthen pit; CC – rectangular, above-ground covered concrete tank.
Apply: I – self reel-in travelling irrigation gun, soil incorporation within 5 days of application; TI – tanker injector.

medium, and large categories. For extra-large hog enterprises, a partially slatted floor replaced the solid floor collection component of the other size categories, otherwise the manure-handling system was identical across the size categories. Note that for economic optimization, no constraints at all were placed on environmental protection needs. For this reason, a number-one ranking for economics generally produced less than optimal rankings for each of the environmental criteria.

This finding can be viewed alternatively by focusing on the optimal solutions under the environmental criteria. Here, the objective was to minimize ammonia (NH_3) emissions from manure operations, or to minimize excess N or excess P in manure applied to cropland, regardless of economic impact (see the second, third, and fourth rows of results, respectively, under each size category, Tables 7.1 and 7.2). In meeting each of these different environmental criteria, there was an adverse economic impact, most noticeably in the case of minimizing NH_3 emissions (second-row results, Tables 7.1 and 7.2). The negative effect on farm net returns was not so large for the other two environmental criteria, but was still noticeable. For example, an operator of a small hog enterprise would have forgone over $2,000 of farm net returns to comply with minimum NH_3 loss restrictions, nearly $1,700 of farm net returns in order to minimize excess N applications, and less than $800 of farm net returns to minimize excess P applications (Table 7.1). These economic costs of environmental compliance would have necessarily been larger for the larger hog-enterprise size categories, as all activities are scaled up.

Besides the economic/environmental trade-offs, some interesting differences in selected feed/manure-handling components were evident among the solutions for different environmental criteria. The optimal feed/manure system for minimizing NH_3 emissions was consistently feed ration 2 with fully slatted floor collection, covered concrete storage, and tanker-injector for field application across all four hog enterprise sizes (second-row results, Tables 7.1 and Table 7.2).

This first-ranked system for NH_3 emissions was ranked 62nd (out of the 81 systems) for excess N applications. In order to minimize NH_3 emissions, manurial N retention was maximized throughout all handling stages, but this automatically increased the N content of manure at field application.

In contrast, for minimizing excess N applied to cropland, manure should be handled so as to encourage NH_3 emissions and reduce N content. This was best accomplished through a combination of a feed ration with lysine, solid floor collection, covered concrete tank storage, and irrigation for field application (third-row results under each size category, Tables 7.1 and 7.2), for all four enterprise sizes. The lack of consistency in optimal feed/manure-handling systems components depending on whether NH_3 emissions or excess N applications are to be minimized is bound to present a dilemma to farm operators directed to reduce both NH_3 emissions and excess N. For all four hog enterprise sizes, the first-ranked system for excess N was ranked 58th for NH_3 emissions. Extra-large enterprises, for example, would see NH_3 emissions almost doubling from 11,675 kg to 22,717 kg as excess N applied declined from 12,572 kg (for optimal NH_3 loss emission solution) to 3,516 kg (optimal excess N solution) (compare second- and third-row results for extra-large enterprises, Table 7.2).

Optimal results for minimizing excess P applied to cropland reflected some peculiarities of manurial P. Regardless of collection, storage, and field application system components, manurial P content would remain the same. For all handling systems examined in this study, complete "watertightness" was assumed so that no manurial P would leak into ground or surface water until after the point of field application. Nor are aerial losses of P possible. The hog farmer is therefore left with feed ration inputs as the only way to manipulate excess P applied to cropland. Hence, only three rankings are considered for feed/manure systems with respect to excess P. First-ranked is feed ration 3, containing exogenous phytase, together with solid floor collection, earthen pit storage, and irrigation for field application (for small, medium, and large hog enterprises) or with partially slatted floor collection, earthen pit storage, and irrigation for field application (for extra-large hog enterprises) (see fourth-row results, all size categories, Tables 7.1 and 7.2). Note that these manure-handling system components were identical to those selected for the economically optimal solution (see first-row results, all size categories, Tables 7.1 and 7.2). In comparing first-ranked excess P solutions with first-ranked economic solutions, while excess P declined some 53% for all hog enterprise sizes, farm net returns declined by only 3%, again across all four sizes. For example, for large enterprises, excess P was reduced from 1,023 kg for feed ration 2 to only 473 kg for feed ration 3; this was achieved at the expense of a fall in farm net returns of $3,937, from $123,600 for feed ration 2 to $119,663 for feed ration 3 (see fourth- versus first-row results, large enterprise, Table 7.2). Thus, compliance with environmental P regulations would not cause as much economic stress for hog-finishing operations as would compliance with NH_3 emission or excess N regulations.

For completeness, we point out that feed ration 2 with synthetic lysine can reduce excess P applied to cropland by 21% compared with feed ration 1 with no added lysine or phytase. Feed ration 3 with exogenous phytase can reduce excess P by 63% compared with feed ration 1.

Conclusions and Implications

From the above research findings, we concluded that:

- There are multiple options for livestock producers to choose from for feeding their livestock and/or handling livestock manure, with each alternative feed/manure-handling combination having different implications for both on-farm economics and society-wide environmental protection.

- Mixed integer programming (MIP) models are useful for identifying the best way to feed livestock and/or handle livestock manure, where "best" can be defined in an economically sustainable (maximum farm net returns) or environmentally sustainable (minimum NH_3, or excess N applied, or excess P applied) context.
- Considerable differences exist between the most sustainable way to feed livestock and handle manure economically and the most sustainable combination environmentally, as measured by farm net returns versus levels of manurial NH_3 gaseous emissions and excess manurial N and P applied to cropland, with environmental goals being achieved only at the expense of economic goals.
- Differences also exist among optimal environmental preparations for livestock feeding/manure handling according to whether attention is directed towards NH_3 emissions or excess N applied to cropland or excess P applied to cropland.
- Addition of synthetic lysine to hog-finishing feed rations for suppression of manurial N and P would be both economically and environmentally advantageous, whereas addition of exogenous phytase for suppression of manurial P would lead to a favourable environmental outcome at only a relatively small expense economically.

These conclusions imply, first, that hog-finishing producers face a dilemma in that farm business economic goals or environmental protection goals can be pursued but not both simultaneously: care of the environment by hog farmers can be achieved only at some cost to farm business profitability. Second, producers face a further dilemma in that different hog-feeding/manure-handling prescriptions are called for depending on which specific environmental pollutant is the focus of attention: a single prescription that simultaneously minimizes NH_3 emissions, excess N applied to cropland, and excess P applied to cropland does not exist. Third, because there is some cost to hog producers in protecting the environment to the benefit of society at large, consideration could be given to offering financial assistance out of the public purse for farmers' environmental protection efforts. This would imply a need for government intervention to ensure that any public subsidies are used for the intended purpose.

While many previous studies have modelled the technical, economic, and environmental trade-offs associated with manure management, little if any work has been done to evaluate manure options in a whole-farm setting. Nor has much work been done in assessing economic and environmental sustainability criteria jointly. The importance of using a holistic, whole-farm approach to manure systems evaluations lies in showing the extent to which overall farm profitability can be affected by environmentally sustainable manure management decisions, the need for government involvement in helping to effect environmentally sound manure operations, and the likely need for financial support for farmers.

8

Balancing Environmental and Economic Concerns in Manure Management by Use of an On-Farm Computerized Decision Support Program, MCLONE4

John R. Ogilvie, Dean A. Barry, Michael J. Goss, and D.P. Stonehouse

Manure management practices in Ontario are becoming of more concern to municipal governments because of potential environmental impacts. Farmers need assistance in designing manure management systems that have acceptable levels of environmental risks and economic costs. In May 1999, an enhanced version of a decision support program for manure management, MCLONE4, was released in Ontario (Stonehouse and Goss 1999).[1] It is written in Visual C++, uses a modern Windows interface, and operates under Windows 9X or NT. MCLONE4 considers the manure management process beginning at feed inputs and animal numbers through field application and losses to the environment. The acronym MCLONE is derived from the word "manure" and the system characteristics that the program evaluates: cost, labour, odour, nutrient availability, and environmental risk.

MCLONE4 combines scientific understanding from several disciplines into a manure management decision-making tool that farmers can apply to their own situations. It was developed from an existing program, MCLONE3 (Ma and Ogilvie 1998) by a multidisciplinary team of experts in soils, crops, economics, engineering, animal science, and atmospheric science at the University of Guelph. Dairy and swine producers in Ontario carried out the alpha and beta testing of the software. Funding was provided by the provincial and federal ministries of agriculture and by dairy and swine producer associations in Ontario. MCLONE4 can evaluate solid or liquid dairy and swine manure systems and solid poultry manure systems, and includes weather, crop, soil, and economic data for different regions of Ontario.

MCLONE4 is also being merged with NMAN99 nutrient management software produced by the Ontario Ministry of Agriculture and Food (OMAF) for use in Ontario. An NMAN99 nutrient management plan approved by

---

1 The software, on CD-ROM, can be purchased from the Canadian Farm Business Management Council, Suite 300, 880 Wellington Street, Ottawa, ON K1R 6K7, for Cdn$60; CFBMC, <http://www/farmcentre.com/english/index/htm> (retrieved 15 July 2004).

OMAF is required by some municipalities in Ontario before they will allow new or expanded livestock production facilities. Some municipalities also require compliance with minimum distance separation (MDS) guidelines for odour. The MDS is the minimum acceptable distance between the livestock buildings and a neighbouring land use that might be negatively impacted by odour from the buildings. The merged program will include manure storage sizing and field application recommendation functions from NMAN99, which are very similar to those in MCLONE4, and the economic and environmental assessment functions of MCLONE4.

This chapter shows how MCLONE4 can provide the farmer with scientifically based information for deciding on a manure management strategy. The main functions of the program are described, and an example is given of its prediction of relationships between various manure application goals and economic benefits. A series of MCLONE4 outputs are provided as a hypothetical example of how the program can be used to design an economical and environmentally sustainable manure system.

MCLONE4 Operation

MCLONE4 uses a main menu screen where the user chooses one of five MCLONE4 functions. The functions are listed sequentially and each can be executed only after all those preceding it have been completed. The functions are:

- estimate manure production and evaluate storage
- estimate crop nutrient needs
- recommend manure application
- estimate manure system costs
- calculate system ratings and environmental risks.

Each function includes screens where the user can input data, and ends with an output screen of results. The user may then continue to the next function or go back to an earlier function and change an input value. This operational flow of the program provides the user (farmer) with feedback at each of the five steps involved in a manure system evaluation (Figure 8.1). Some outputs from each function are used as inputs for subsequent functions.

The first function estimates manure production based on animal numbers and types and, if available, feed amounts and composition. It requests the type and dimensions of manure storage structures and the duration and season of storage. It assesses the adequacy of storage capacity, including inputs of bedding, wash water, and precipitation. Information is also requested about the dimensions of the barn and distance to nearest neighbour, for use later in the program to calculate odour dispersion. The function output summarizes all the inputs to storage, ammonia (NH_3) losses from

Figure 8.1

MCLONE4 operation, functions, inputs, and outputs

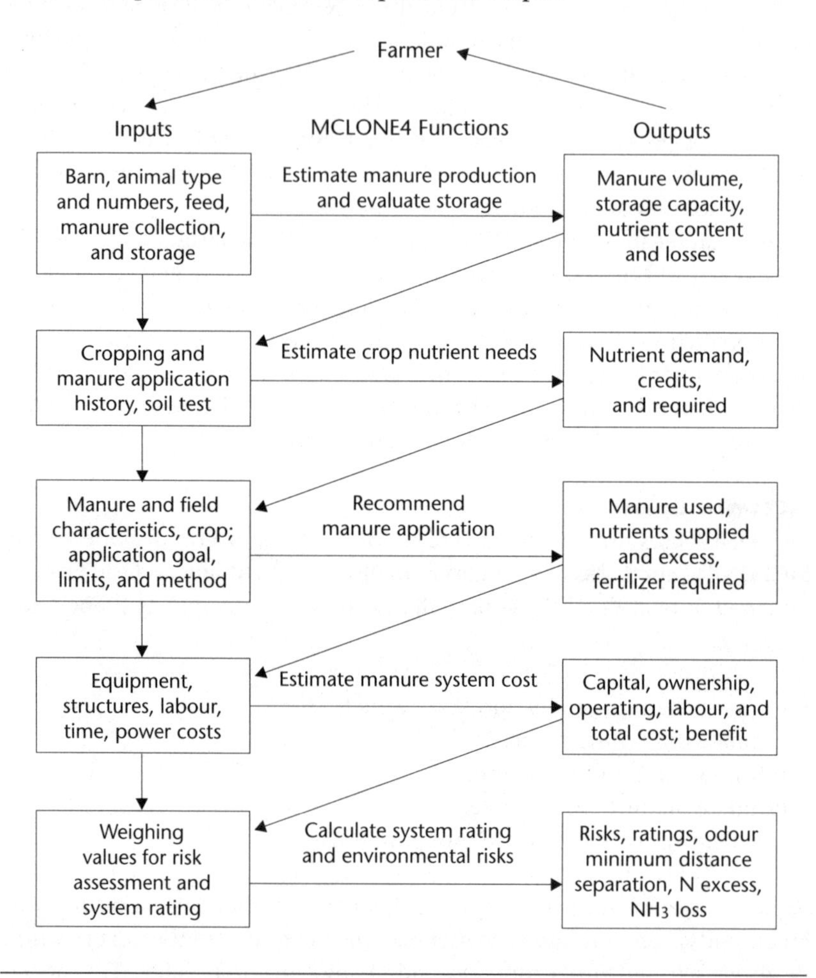

the barn and storage, and final manure volume and composition, and alerts the user to any shortfall in storage capacity.

The second function estimates the nutrient requirements of the crop by using the standard field crop recommendations for Ontario. These recommendations have been developed from empirical relationships observed in field trials conducted on farms and research station plots throughout the province over the last several decades. The function inputs include crop type, expected yield, geographic region, previous crop and manure applications, and soil test levels for phosphorus (P) and potassium (K). Additional information about the field (slope, soil texture, residue cover, planting and application date, soil moisture, tile drainage) and tillage practice are requested for later calculations of potential nutrient losses by runoff, leaching, and volatilization. The output indicates the amounts of plant-available nitrogen (N), P, and K required from fertilizer or manure application, considering available nutrients from soil and N credits from the previous crop and previous two years' manure applications.

The third function provides a recommendation regarding the amount of manure and additional fertilizer to apply for the crop. The user must indicate the method of application and labour requirement, time to incorporation, and application goal for each field where manure will be applied. Five choices are provided for the application goal:

1 meet 75% of the crop N requirement
2 meet 100% of the crop N requirement
3 meet 100% of the crop P requirement
4 apply the maximum amount of manure (number 2 or 3, whichever is greater)
5 a user specified rate.

The recommended rate will be the least rate of three possibilities: the application goal, a hydraulic loading limit for the soil, and a user-specified maximum allowable rate. The hydraulic loading limit is the maximum rate of liquid manure that can be applied without causing runoff, and it is calculated by the program. The function output summary gives the three possible rates, the recommended rate, the amount of manure that would be used and amount remaining, the area (size) of the field that would receive manure, and the area remaining if all the manure is used. It also indicates the amount of N, P, and K supplied by manure and the remaining nutrient needs that should be provided by fertilizer. If the manure rate is in excess of crop nutrient requirements, then the amounts of excess nutrients are indicated. For P, a warning message indicates the number of years until the annual manure application causes the soil P level to exceed a maximum acceptable value.

The function to estimate manure system cost uses a combination of user input values and default values to estimate four categories of cost: capital cost, ownership cost, operating cost, and labour cost (Table 8.1). These costs are determined for the manure collection, storage, and application systems separately. The user must provide sufficient details for the program to estimate the costs of all structures and equipment involved in the manure system, including the proportion of time that equipment, such as a tractor, is used for manure handling. The capital cost is the purchase price for structures and manure-handling equipment. The program will estimate these costs as an average retail price according to specifications and capacity of each type of equipment, by using a database of prices. The user can see a

Table 8.1

Cost information from MCLONE4 output of the estimate cost function

Floor type .... Dairy free stall, warm barn, with slatted floor
Collection method .... Gravity
Total manure per year .... 1,553.5 m^3

System component	Initial capital investment ($)	Annual manure system costs Ownership ($/yr)	Operating ($/yr)	Labour ($/yr)	Total annual costs ($/$m^3$)
Floor and under barn*	49,084	7,315	818		5.24
Collection	0	0	228	1,825	1.32
Open circular tank	23,976	2,801	400		2.06
Storage pump/agitator†	14,993	2,234	498		1.76
Tanker broadcast†	27,778	4,140	873	258	3.39
Total cost	115,831	16,490	2,817	2,083	13.77
Benefit of manure			($6,278)		(4.04)
System cost					9.73

* Under-floor storage, includes additional costs for the manure system only.
† Includes a percentage of the cost of tractors used to operate the equipment.

summary screen of the estimates of capital costs for their structures and equipment and can change the costs. These values are used in determining the annual ownership costs for the manure system components. The ownership cost is an annual amortization cost that depends on an opportunity cost interest rate and the expected life of the facility or equipment. Operating costs include energy, lubricants, bedding, maintenance and repair (10% of capital cost), and insurance (1% of capital cost). The labour cost is estimated from labour hours for manure handling and an opportunity cost of labour, both estimated by the farmer.

Upon completion of the estimate cost function, the output screen provides total annual costs divided by the annual volume of excreted manure, for the purpose of comparing different systems (Table 8.1). This manure volume does not include additional inputs such as bedding, wash water, and precipitation. The output also reports the economic benefit of the manure and deducts this value from the total cost to give a net system cost. The manure economic benefit is based on the nutrient benefit of the manure to the crop in terms of its replacement cost by fertilizer N, P, and K. Only the plant-available portion (equivalent to manufactured fertilizer) of applied manure nutrients up to the level of the crop's nutrient requirements are included as a benefit. It is assumed that the proportions of manure nutrients that are as available as manufactured fertilizer in the year of application are 100% of the inorganic N, 30% of the organic N, 40% of the total P,

Figure 8.2

MCLONE4 predictions of manure economic benefit and excess nutrients

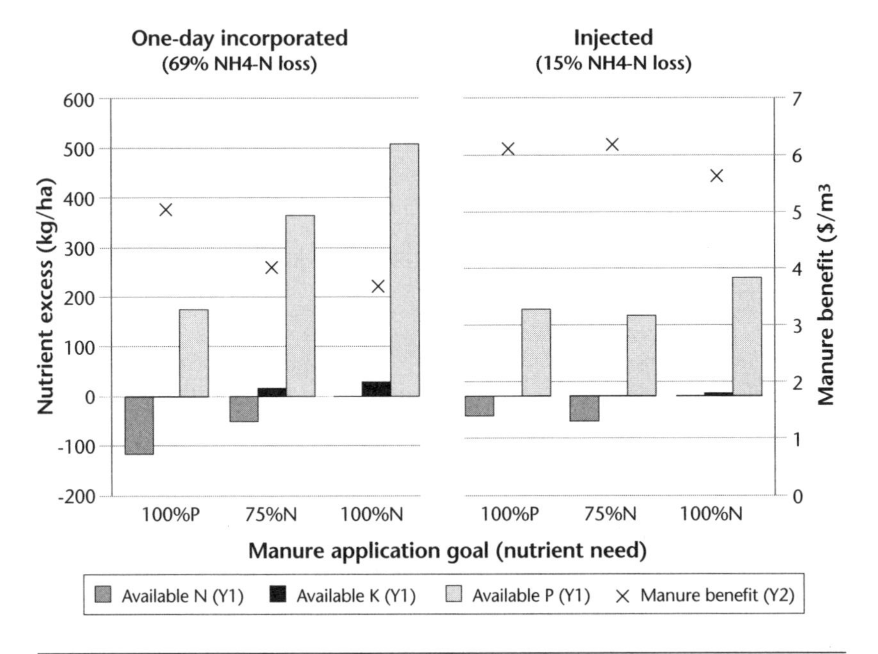

and 90% of the total K. Loss of N by volatilization during and after application is accounted for in the calculation of required application rate and manure benefit.

An example of the relationship between manure economic benefit and application goal is given in Figure 8.2 for two types of application of liquid dairy manure: (1) broadcast by tanker and then incorporation in one day, and (2) injection from a tanker.

Application is assumed to be in the spring and for a corn crop. The soil is assumed to have medium levels of available P and K. The additional amounts of available nutrients required for the corn crop and to be added as manure or fertilizer are 200 kg N per ha, 22 kg P per ha, and 66 kg K per ha. For the broadcast system, manure application rate increases from 75,440 L/ha to meet 100% of the P requirement, to 134,434 L/ha to meet 75% of the N requirement, and 179,245 L/ha to meet 100% of the N requirement. The manure benefit decreases as application rate increases beyond the 100% P goal because of oversupply of P and K. For the injection system, the application rate required for the 100% P goal is the same as with the broadcast system, but the manure benefit is greater with the injection system. The greater manure benefit results from the decreased gaseous loss of N. The consequently greater available N supply with injection also results in a better balance between available and required nutrients from the manure, so oversupply of P and K when application rate is based on the N requirement is less with injection than with broadcast application.

The environmental risk and system rating function involves calculating a risk level (low, medium, or high) for each of the five environmental risks:

- odour level at the nearest neighbour
- NH_3 loss
- P loss by runoff
- nitrate leaching
- bacteria transport.

Criteria for assigning risk levels are listed in Table 8.2. The risk levels are also given numerical values of 1 for high, 2 for medium, and 3 for low risk,

Table 8.2

Criteria for assigning ratings to environmental risks of a manure system

Criteria	Low risk	Medium risk	High risk
Odour Production and Dispersion[a]			
Odour level at nearest neighbour	<2 * threshold	2 to 15 * threshold	>15 * threshold
Ammonia Loss by Volatilization[b]			
Barn: air temperature, floor type	< 8°C, any	≥ 8°C, slatted	≥ 8°C, solid
Storage: NH_4-N loss	< 11 %	11-50 %	> 50 %
Field: NH_4-N loss	< 30 %	30-60 %	> 60 %
Phosphorus Loss by Field Runoff[c]			
P index	< 15	15 - 30	> 30
Nitrate Loss by Leaching Past the Rooting Zone[d]			
F1 * F2	≤ 2	4	≥ 8
Bacterial Contamination of Ground Water[e]			
$\log_2$ (X1*X2*X3*X4*X5)	≤ 3	4 or 5	≥ 6

a The threshold odour level is an assumed maximum acceptable value derived from Ontario minimum distance separation guidelines.

b Reported risk is the maximum of barn, storage, and field risks.

c P index is based on soil erosion, soil runoff class, soil test P, P source, application rate and method, and proximity to surface water. Reported risk is the maximum of all fields.

d F1 is a field N balance rating; 1, 2, or 4 for <7.5, 7.5-15, and >15 kg excess N ha-1, respectively. F2 is a deep percolation rating; 1, 2, or 4 for <30, 30-60, and >60 mm of percolation water, respectively. Reported risk is the maximum of any combination of field and season.

e X1...X5 are ratings for depth to ground water, soil moisture status, season, manure type, and manure incorporation.

for calculating a weighted average environmental risk value. This risk is transformed to a scale of 1-10 as used for the manure system ratings for cost, labour, odour, nutrient retention against loss to the environment, and environmental risk. The 1-10 scale (poor to excellent) is based on the worst and best possible systems available in the program. The system odour rating and the odour risk rating are both based on a worst-case scenario for odour dispersion. The odour risk is based on the estimated odour level at the nearest neighbour, while the system odour rating is based on the distance required to obtain proper odour dilution (a minimum distance separation). The system rating and risk rating use inputs and results from the previous four MCLONE4 functions. The only additional input required is choice of weighting values for each of the five risk and five system rating components. The weighting values for five components must sum up to 100. The program combines these values with the individual system and environment ratings to calculate an overall environmental risk rating and an overall system rating that consider the individual farmer's priorities for manure management.

An Example Scenario of MCLONE4 Use

Comparison of outputs for a series of modifications to a liquid manure system on a dairy farm is used to show how MCLONE4 can be used to balance economic costs against environmental risks when designing a manure system (Table 8.3). The output is from the system rating function, the ammonia loss summary, and the total annual costs component of the cost function. The farmer would begin by working through the first four MCLONE4 functions to ensure that the output from each function indicates a potentially workable system. The ratings would then be used to suggest areas for improvement.

The ratings for an initial manure management system (system 1 in Table 8.3) indicate a high risk of ammonia loss. The output information on ammonia loss indicates that most of the loss occurs in the field, and so is a consequence of broadcast application with incorporation about one day later. The farmer might therefore decide to change to an injection application system (system 2). This choice decreases the ammonia volatilization risk but results in a high nitrate leaching risk. The risk occurs because the amount of N susceptible to leaching has increased (due to conservation of manure inorganic N) and a significant amount of deep percolation can occur because of the early spring application (Table 8.2). One option is to delay manure application until late spring, when there is less deep percolation. This option (system 3) results in low to medium environmental risks, so the farmer might regard these as acceptable and examine the overall system ratings.

The poorest system rating is for cost, which is average but only slightly better than poor. The total annual cost was relatively high for the under-barn

Table 8.3

Environmental and economic results for liquid dairy-manure systems in southwestern Ontario

Rating criteria	Manure management system[a] 1 Slatted floor, concrete tank, April broadcast, 1-day incorporation	2 Change (1) to drag-hose injection	3 Change (1) to June tanker injection	4 Change (3) to wide-alley barn	5 Change (4) to lagoon
	Environmental risk index (low, medium, or high)				
Odour	m	m	m	m	h
Ammonia loss	h	l	m	m	m
Phosphorus	l	l	l	l	l
Nitrate leaching	l	h	m	m	m
Bacteria	m	m	m	m	m
	System rating (<3 = poor, 8-10 = excellent)				
Cost	4.7	3.6	3.9	5.7	6.5
Labour	9.0	9.0	8.9	8.9	8.9
Odour	5.0	6.2	6.2	6.2	1.0
Nutrient	7.3	10.0	9.7	8.7	8.7
Environment	6.4	6.4	6.4	6.4	5.5
System	6.5	7.1	7.0	7.2	6.1
	Environmental risk information				
Odour MDS (m)	176	176	176	176	329
Barn NH_3 loss (%)	10	10	10	18	18
Storage NH_3 loss (%)	8	8	17	17	17
Field NH_3 loss (%)	69	15	15	15	15
N leaching (kg N/ha)	0	16	37	37	37
Deep percolation (mm)	37	37	7	7	7
	Cost summary ($/m³)				
Floor and under barn	5.24	5.24	5.24	0.44	0.44
Collection	1.32	1.32	1.35	3.14	3.14
Storage	2.06	2.06	2.06	2.06	0.74
Application	5.15	7.02	6.44	6.44	6.52
Manure benefit[b]	4.04	6.18	5.69	5.18	5.18
System net cost	9.73	9.46	9.40	6.90	5.65

a Some assumptions for the manure system were 600 m to nearest neighbour, 54 dairy cows, 200 days manure storage capacity, manure application to a 20 ha corn field to meet 75% of the crop N requirement, expected corn yield 10 t/ha, loam soil, 1% slope.

b Equivalent synthetic fertilizer value, not inflated by cost of spreading such fertilizer.

storage because of the slatted floor system. The farmer might consider a wide-alley barn. This choice decreases the floor and under-barn cost by an order of magnitude (system 4). Now the cost rating is average and nearly good. Another possible cost-saving measure would be to use an earthen lagoon rather than a concrete tank as the main manure storage. This system (5) results in a high odour risk by nearly doubling the recommended minimum distance separation, because of the much larger surface area of a lagoon compared with a tank of the same capacity.

Conclusion

The decision support software MCLONE4 assists farm operators in designing a manure management system that suits their own resources and concerns. It enables the farmer to evaluate the entire manure system from animal feed to crop production and losses to the environment. The farmer must have a basic understanding of the biophysical processes considered by the program in order to use the program interactively in decision making.

Design and operation of a manure management system requires simultaneous consideration of environmental and economic outcomes. In 1999 a multidisciplinary research team released MCLONE4 for use by farmers and farm advisors to help in the selection and management of a manure system. The Windows 95/98/NT program requires the user to provide information about animals (dairy, swine, or poultry), manure type (liquid or solid), feed rations, barn construction, manure storage, and field, crop, and application components. MCLONE4 provides manure application rate recommendations and also manure system ratings for cost, labour, odour, nutrient efficiency, and environmental risks. Ratings are weighted according to user priorities to give an overall system rating.

Environmental risks are odour pollution, ammonia loss, phosphorus runoff, nitrate leaching, and bacterial transport to a water resource. These risks are evaluated and averaged according to user-weighted values to give an overall environmental risk rating. The odour risk rating evaluates dispersion of odour for various climatic and topographic conditions and determines the risk of offensive levels of odour reaching the nearest neighbour.

The economic evaluation makes use of an included database of purchase costs for all equipment used in common manure systems. The user can modify manure system components to achieve a better balance of economic versus environmental results. Extensive documentation of program inputs and why they are needed is given in online help.

9

Challenges Awaiting the Dairy Industry as a Result of Its Management Decision Environment

Wayne C. Pfeiffer and Glen C. Filson

In local or micro-economic terms, a notion has been promulgated in North America for at least forty years that farming must be treated "more like a business and less like a lifestyle." As farming becomes more industrialized and evolves towards more intensive land use, concentration of livestock into confined "growth centres," and greater reliance on labour-saving technology, critics have argued that a lack of managerial acumen has brought the system to the brink of becoming an environmental disaster as well as economic paralysis with no further investment incentives.

Prior research convinced us that dairy farming has become one of the most pleasant and satisfying occupations in Ontario agriculture. Nearly 200 randomly selected Grand River dairy farmers recently surveyed rated their quality of life as high and considered themselves better off than the average Canadian. Gone are the days of surplus production, low prices, and the economic chaos that characterized dairying prior to the mid-1960s, when the present supply managed system was introduced.

The agricultural situation throughout the developed and much of the developing world reflects societal attitudes towards food production that stem largely from two sources: (1) the popular thesis espoused by T.R. Malthus in the early nineteenth century that adequacy in food production will always be pressured by population growth, and (2) the "never again" philosophy of participants in the major wars of the twentieth century, referring to dependence on other nations for food. One hundred years of agricultural education and research coupled with fifty years of isolationism and subsidization have resulted in pockets of chronic overproduction and low farm profitability needing public support, and skewed balances of production and distribution that have left pockets of famine and a growing concern that the entire project was not sustainable for both economic and environmental reasons. To say that better management needs to be brought into agriculture's "big picture" is one of the twenty-first century's leading understatements. Perhaps Malthus was correct after all.

Farming Decision Making and Ontario Dairy Farm Sustainability

Because the theme of the research summarized in this book was identification of problems and solution strategies for making modern North American agriculture more sustainable, the socio-economic farming systems research (FSR) team at the University of Guelph sought to investigate the roots of farm decision making. Many farming systems and farm management studies have reported difficulty in explaining why different farm operators controlling essentially similar amounts and quality of resources posted wildly differing degrees of economic success.

In general, this aspect of farming systems research was motivated by the hypothesis that the physical and economic environment on farms that gives rise to farmers' perceptions about their quality of life in turn affects the managerial decisions themselves. If a climate of unhappiness and the perception of low quality of life is dominant, we expect a short-run reactionary, risk-creating palette of attitudes and attendant short-sighted business decisions. Conversely, we expect that in a climate of happiness and perception of high quality of life, the palette of attitudes will include both short- and long-term notions. In such a climate, planning will eclipse knee-jerk reactions to market fluctuations, and long-term risk-taking forms of investment will be the norm.

Sectors of agriculture differ widely in their ability to produce stable business conditions for participants (producers, processors, retailers). Some sectors have low levels of social capital in the form of marketing agencies and sophisticated public sector involvement. These are often found to have high levels of rather simplistic public sector support, the most blatant being market price support at the point of production. By contrast, other agricultural sectors have evolved away from the chaos of volatile markets driven by reactionary production swings. These often show surprising levels of public/private sector interaction, production coordination, and investment. In Ontario, no sectors that fit these criteria stand in greater contrast with each other than pork and dairy.

The dairy sector also possesses the most attractive characteristics for studying the nature of management decisions. In virtually all jurisdictions, dairying has demonstrated a proclivity for overproduction. On the one hand, the dairy industry is highly systematized, largely because of the perishability of the product and secondarily because of the many steps necessary to bring the product to market. Ever since pasteurization of milk began in New York City in 1907, dairy product quality assurance made public sector involvement in the industry the norm. Increasing productivity resulting in supply expansion kept milk prices low, and dairy farmers poor, for decades. Individual farmers' attempts to maintain income through increased output in the face of falling prices, aided by advances in dairy technology (milking machines, selective breeding, nutritional knowledge, and so on) generated a collective impetus towards surpluses for decades. W.W. Cochrane (1958) described the process as the "Agricultural Treadmill," and laid the problem at the feet of two major players: (1) creators of agricultural technology and (2) farmers, whose lack of understanding or constraint of prior commitments prevented them from taking the best course of action in the face of low profitability, namely, contraction of output rather than expansion.

Supply Managed Dairying in Ontario

For decades, public sector involvement beyond enforcing quality standards traditionally resulted in the Band-Aid measures of price supports and surplus removal, until a unique and forward-thinking jurisdiction broke with tradition and ended the "subsidy cycle" in dairying. In the 1965 Milk Act, the Ontario government passed the enabling legislation for the first supply-controlling marketing agency in agriculture. The culmination of decades of wrestling with the dairy problem resulted in a publicly sanctioned system that forced farmers to collectively limit the industry's output to what the market would bear at a price that was calculated by formula to cover the cost of production. Formula pricing, milk pooling, rationalized transportation, point of sale to processing plants, and production quotas were combined in one stroke to produce the most sophisticated set of industry regulations anywhere in the world.

In Ontario the struggle to calm the wild economic conditions in dairying reached noteworthy proportions after the Second World War. Anticipating a continuing scenario of farm milk price supports accompanied by surplus production and regional disparity, the industry and government arrived at a decision to legislate a new institutional structure for the industry. In the new arrangement, it was expected that dairy farmers would collectively have greater market power as well as greater responsibility for managing the economic affairs of the industry. This, of course, was done in addition to the other socially accepted interventions, such as quality inspection and veterinary health standards (Biggs 1990).

The managerial impact of the Ontario system of dairy industry regulation was profound. For example, no longer were farmers at the mercy of the local processor for price, but at the same time no longer could they produce on a seasonal basis. In order for dairy farmers, milk processors, and cheese makers to survive, their managerial strategies – if they had any at the time, were forced to adjust. Pricing agreement, cost-efficiency, and cooperative behaviour were all thrust upon the industry. Eventually the success of the Ontario Milk Marketing Board (OMMB) – renamed the Dairy Farmers of Ontario (DFO) in the 1990s – led to a national supply management system in 1970 (Veeraraghavan 1985).

The Ontario supply management system required that dairy farmers collectively govern their output within the confines of total milk demand. It was hoped that this level, arrived at by historical precedent, would also be scientifically researchable. The new entity (the first supply management marketing board) was also given the legal right to be the sole owner (and seller) of milk onward into the supply chain, bestowing upon it the market power of a monopoly. While the new system greatly increased the bargaining power of milk producers, it was much more than a pricing monopoly. It was legally designated to be the designer and regulator of the milk transportation system, controller of the raw product flow to primary processors, and "paymaster" to the production sector. As a hedge against abusive wielding of this power, the government retained a veto over any proposed increases to the price of milk. To assist decision making in such an environment, a complex formulaic process was devised to keep everyone apprised of the cost of producing milk, and therefore of the price farmers needed in order to earn a normal profit. This formula pricing system quickly became the subject of controversy that continues to this day.

In order to keep total milk output from expanding too fast under favourable pricing conditions, the new entity was also given the power to enforce milk production quotas on individual farms. Finally, it was also given punitive power to rescind quotas in the event that they were not filled regularly throughout the year.

Advocates of supply management typically cited higher and more stable farm gate milk prices, improved efficiency in transporting milk from farm to processor, reduced seasonality of supply, less dependence on subsidy, elimination of surpluses, and greater public awareness of the industry's needs in maintaining a steady supply of safe, high-quality products. Mostly the arguments mounted by advocates either compared current conditions with those that prevailed in earlier times, or had a short-term viewpoint, or both. Organized business thrusts have recently been made by milk producers who do not hold production quota to operate outside the system by shipping directly to export markets, posing a potential new threat to supply management. Also, the General Agreement on Tariffs and Trade (GATT) led to the creation of the World Trade Organization (WTO) in 1995. The upshot of all this is that the system of supply management in the dairy industry has come under renewed scrutiny.

Detractors usually cited such futuristic things as sluggish adoption of efficiency-raising improvements on farms over time. This, they argued, would keep the cost of producing milk higher than would otherwise be the case. They also warned of the high cost of expansion to meet a growing population's demand for milk, brought about by the high price for production quota units in the quota exchange market. This, they expected, would thwart the expansion of small production units that had limited capital with which to purchase quotas in competition with larger farms. Finally, they predicted erosion of consumer confidence in the industry's ability to recognize market signals on the demand side, resulting in inflation of store prices for milk over time. Most of the detractors' arguments were futuristic, with a predictive viewpoint.

On 24 June 2002, the WTO came out with a ruling "which says Canada is breaking global trade rules through its dairy export system, confirm[ing] the arguments of international critics who have complained that dairy processors can buy milk for export products at steep discounts" (Chase 2002).

The supply management system attempts to make all dairy production appear to the supply chain as coming from one monopolistic producer. As such this represents an example of industrial organization backed by regulatory fiat. The WTO has now said, however, that giving the dairy industry that degree of regulatory power is itself a subsidy. This argument is countered with the reminder that the government retains approval over any changes to the price of milk and attempts to strike a balance between consumers and the milk-producing sector by maintaining a statistically defensible cost of production formula on which to base pricing decisions in the absence of unfettered forces of supply and demand. At present, the old term "supply management" is being superseded in the dairy industry by the term "orderly marketing," and the industry argues that the outcome in terms of milk price is not substantially different from what the free market forces would determine.

The current decision in the WTO is to allow this procedure under amber lights; however, the WTO cites tariff barriers that still exist as needing to be reduced substantially so as to open the region of orderly marketing to the supply and demand forces of the world milk market. The livelihoods of a small number (eighty) of dairy farmers who have no domestic quota but instead depend entirely on the export market are threatened by the WTO ruling that Canadian producers are receiving an unfair subsidy through the DFO and related agencies across Canada. The DFO chair has ruled that, given the WTO decision, there is no room for the separate export system (Whalen 2003).

We believe that in order to judge appropriately the arguments for and against orderly marketing, the impacts on the lives of the managers in the dairy industry need documentation. The quality-of-life study was designed to clarify how dairy farmers perceived their sense of well-being, and the environmental and economic issues they faced, including their views about how they thought of such things as the North American Free Trade Agreement (NAFTA) and GATT when the survey was conducted just before GATT evolved into the WTO.

In order to ensure that sufficient variability in major indicators of quality of life was present, our study sought a geographic area that harboured a wide range of dairy farming characteristics. Such a study area would exhibit a sufficiently long dairying history to encompass a wide demographic cross-section. It would also exhibit wide variability in biophysical factors to ensure that environmental considerations were likely to be present in the mix of farm decisions. Finally, the chosen area would occupy a land base on which various viable sources of competition exist in the form of alternative agricultural enterprises.

In southwestern Ontario, the watershed area of the Grand River lies midway between Toronto and London. Within this watershed is an agriculturally diverse area with urban development and environmental fragility sufficient to ensure a rich mixture of the underlying components for a study of farmers' perceptions and how they may correlate with farm profitability, environmental awareness and stewardship, public interventions, capital investment, and the overall decision-making climate.

Our survey of a representative sample of 194 Grand River Watershed dairy farmers revealed many farmers who expressed concern that they would have to expand their operations or get out of dairying, in part because economists' predictions of soaring production and plummeting prices appear to be coming true in many agricultural sectors. Responses from these farmers also showed that while one of the outstanding concerns of producers is the price of milk (which is higher in Canada than in the US due to supply management), the reason for the concern that many had about their relatively small herd sizes extends well beyond a simple calculation of profitability. As they report, profitability is considered to be a necessary condition for staying in the industry but not a sufficient condition by itself.

Our farm interviews, focus groups, and farmer questionnaires with these dairy farmers showed that their farms usually had many of the hallmarks of sustainability, including good productivity, excellent viability, stability, and especially social acceptability. This was due to their steady supply managed incomes, excellent cattle genetics, and strong family and church/community relationships. Somewhat weaker, on average, was their degree of environmental protection. At least a third felt that they had manure management systems that were inadequate in the face of attempts to expand their herds, assuming they could afford to buy the increased milk quota (Filson et al. 2003).

In the survey results, there are clear differences between age groups. Middle-aged farmers comprise the largest group and have the largest herds on average. Their farms also display the widest range in herd size. Both of these facts were expected because it takes time to build a large enterprise and the largest age category is most likely to exhibit the greatest diversity.

Whereas the latter was expected, it is interesting to note that the majority of larger-than-average herds belong to younger-than-average farmers. Are we seeing the beginning of a concentration into larger herds? Certainly, the number of Ontario dairy farms has declined from 404,797 to 363,544 (by 10.2%) between 1996 and 2001 alone,[1] with the move towards fewer but larger farms. Some say that our present direction will lead ultimately towards the phenomenon in dairying that has been called the "disappearing middle" in the United States. In this scenario, farms with large herds drive those with medium-sized herds out of business, which leaves a "hole" in the middle of the farm size profile (Lyson and Gillespie 1995; Schwarzweller 1996). Although the freer market approach characteristic of most American dairying is tending to create a bimodal distribution of relatively small family farms on the one hand and relatively large, corporate and even factory-type dairy farms on the other, the supply managed Ontario dairy farms have nevertheless generated more of a normal curve, which enables those with a medium number of milking cows to thrive (Filson et al. 1998).

For the "disappearing middle" phenomenon to happen, it is necessary to have either a great number of farmers old enough to retire whose average herd size is small or farmers who have recently expanded into medium-sized herds who are in financial straits and wish to undo their mistakes by leaving the industry or downsizing. Ontario dairying has both of these groups and both of them clearly express concern for the future, judging from their responses to our questions about government involvement in the industry and potential policy changes.

In this part of Ontario, dairy farmers around the age of forty-five years expressed the greatest concern for the future, and also listed hard work and long hours as major factors of daily stress. The actions taken by this group are likely to be key determinants of the profile of dairy farming for many years to come (Filson et al. 2003).

The farms belonging to this middle-aged group also exhibited the greatest diversity in both herd size and efficiency. From their ranks will come the greatest pressures resisting policy change. Those who are already planning an exit from the industry are very concerned about quota values and the potential impact of the WTO and NAFTA. Evidence can be found that while overall they consider their quality of life to be quite good now, they expect that it may be diminished significantly if the production quota system were to change. This group will likely mount pressure to preserve the present system intact. Because of sheer numbers, their voice will be heard and will likely sway attitudes throughout the industry; already a slight note of cynicism can be heard. They reported lower than average incomes per cow and exhibited extreme concern for milk prices. Interestingly, they also expressed the greatest pride in their herd's performance than any other age group in the survey (Filson et al. 2003).

---

1 Statistics Canada, "Cattle and calves, provinces," online at <http://www.statcan.ca/english/Pgdb/econ105g.htm> (retrieved 19 June 2002).

Research Questions

Now that the rules of the game are changing (because of adjustments being made to accommodate Canada's inclusion in the WTO) and new players are taking the field, what lies in the future of dairy farming in Ontario? Any accurate prediction of change in the dairy industry not only must be based on the study of the history of change in the industry but must also consider people's attitudes towards work, the environment, family, money, social status, other people, politics – in short, commercial agriculture's survival rests as much with human factors as it does with biophysical factors.

For the first time since the 1960s, when the Milk Act created the present structure of the industry, young people are entering the milk-producing sector in significant numbers. In fact, the average age of dairy farm operators has fallen by a decade in just five years. Answers to questions arising from changing conditions are vital to understanding the change that is rapidly accelerating in dairying today. What do the new young dairy farmers perceive will be the future of the industry? Will they pursue different personal and business goals than their elders?

The milk market in Ontario today involves both restrictions and protection. Can it survive in the changing political climate of agriculture? How will it change? How will dairy farming change if the marketing system changes? Having a profitable operation is obviously a motivator in dairying. But does that mean that once profitability is achieved at a certain herd size, expansion will follow?

The 1997 Survey

In 1997 a sixteen-page survey was sent to 465 Grand River dairy farmers out of a total number of about 1,600. It was completed by 194 dairy farmers (Table 9.1), or 42% of those who received the mailed questions about their farming operation, family, and quality of life.

Over 80% of their mean income came from dairy farming. Relative to other agricultural commodities, dairying has the least resort to off-farm work, and this partly accounts for the farmers' relatively higher perceived quality of life (McCoy and Filson 1996).

The mean number of full-time employees on dairy farms was 0.4 and the mean number of part-time year-round workers was 0.325, with an average of 0.7 casual workers. Farm size was positively correlated with the number of part-time employees ($r = .242$, $N = 191$, $p \leq .000$). Custom operators and hourly employees were utilized most for harvesting, planting, silo filling, manure spreading, milking, and feeding, in that order.

An average of 1.4 families (each comprising about 5.7 family members) were reported to be dependent on each farm's income (Pfeiffer and Filson 1999). Net farm income was somewhat positively correlated with the size of their farms, their herd's performance, and their number of milking cows

(Table 9.2). The vast majority (87.3%) said that the size of their income was either important, very important, or extremely important to their quality of life. Farm size and net farm income were therefore associated, although there are farms with small acreages earning more than $160,000 and also large farms earning less than $50,000 (Table 9.3).

Table 9.1

Demographic characteristics of the respondents to the 1997 survey of Grand River Watershed dairy farmers

Characteristic	Number	Percent	Average (years)	Range (years)
Age group			46.5	23-79
Less than 35 years	35	18.0		
36-45 years	67	34.5		
46-54 years	49	25.3		
55 years and over	43	22.2		
Gender				
Male	177	91.2		
Female	17	8.8		
Education				
Primary school or less	71	36.8		
High school	70	36.3		
Post-secondary	52	26.9		
Those living on their farm	183	94.3		
Number of years farming			24.4	1-80
9 years or less	27	14.0		
10-19 years	49	25.4		
20-29 years	47	24.3		
30-39 years	45	23.3		
40 years or more	25	13.0		
Number of years farming in present location			21.3	1-80
9 years or less	39	20.2		
10-19 years	58	30.1		
20-29 years	40	20.7		
30-39 years	38	19.7		
40 years or more	18	9.3		
Number of years farm has been in family			46.4	1-190
24 years or less	52	29.2		
25-49 years	61	34.2		
50-99 years	49	27.5		
100 years or more	16	9.0		

Table 9.2

Correlations between net farm income and farm size, herd performance, and number of milking cows based on the 1997 survey of Grand River Watershed dairy farmers

	Net farm income	*N*	*p*
Farm size	$r = .196$	191	.008
Herd performance	$r = .195$	183	.009
Number of milking cows	$r = .156$	191	.036

Table 9.3

The association between farm size and net farm income based on the 1997 survey of Grand River Watershed dairy farmers

	Net farm income *n* (%)				
Farm size	Less than $49,999	$50,000-$79,999	$80,000-$119,999	$120,000-$159,000	$160,000 or more
40.4 ha or less	29 (49.2)	11 (18.6)	7 (11.9)	5 (8.5)	7 (11.9)
40.5-60.7 ha	8 (22.2)	7 (19.4)	9 (25.0)	6 (16.7)	6 (16.7)
60.8-80.9 ha	13 (30.2)	8 (18.6)	10 (23.3)	4 (9.3)	8 (18.6)
81 ha or more	9 (19.6)	11 (23.9)	5 (10.9)	6 (13.0)	15 (32.6)

$\chi^2 = 21.07$; df = 12, $p \leq .049$

Average herd size was 93.1 cattle, ranging from 23 to 371 animals. Milking cows made up most of these herds (average, 45 cows).[2] Inspection of the differences between those with medium and large numbers of cows also shows that, on average, a slightly higher percentage of those with between 30 and 49 milking cows had larger net incomes than those with 50 or more (Table 9.4).

Table 9.5 shows the importance of the operating conditions that influenced dairy farmers' perceived quality of life the most and the least. The importance of the quota/supply management system to these farmers is shown by these data, followed by the economic health of their community, then their costs of production. These dairy farmers were then asked to compare their quality of life with that of other people. Table 9.6 presents these results.

Thus, these Grand River dairy farmers compared their perceived quality of life very favourably with that of American dairy farmers, urban Ontarians, other Ontario farmers, and Canadians in general. Weersink, Nicholson, and

2 This compares with an average of 55 for the province for 1998, for the Ontario Dairy Farm Accounting Project's sample (ODFAP 1999).

Table 9.4

Number of farms by total net farm income and size of milking herd based on the 1997 survey of Grand River Watershed dairy farmers

	Size of milking herd			Total number
Net farm income	29 or fewer (%)	30 to 49 (%)	50 or more (%)	of farms (%)
Less than $49,999	20 (33.9)	23 (39.0)	16 (27.1)	59 (32.1)
$50,000-$119,999	11 (16.2)	25 (36.8)	32 (47.1)	68 (37.0)
$120,000 or more	7 (12.3)	30 (52.6)	20 (35.1)	57 (31.0)
Total	38 (20.8)	78 (42.8)	68 (36.4)	184 (100.0)

$\chi^2 = 13.16$; df = 6; $p \leq .011$

Table 9.5

Operating conditions affecting Grand River Watershed dairy farmers' perceived quality of life

Factors affecting quality of life	Mean response[a] (0 to 4)	% indicating very positive[b]
Quota/supply management system	3.20	48.0
Economic health of their community	2.64	9.0
Costs of production	2.19	10.3
Getting and keeping hired labour	2.16	10.5
Need to comply with government regulations	1.95	1.8
Availability of rental land	1.69	5.5
Property taxes	1.66	4.6
Outflow of rural youth to the city	1.63	4.7
Urban expansion to rural areas	1.20	2.9

a Responses ranged from "very negative" (0) to "very positive" (4).
b Percentage of respondents whose answer was "very positive" (4).

Weerhewa (1998) argued that the higher level of security and economic well-being perceived by Ontario dairy farm families was likely to be eroded if economic liberalization took place and farms started to operate in a more "competitive" environment. While trade liberalization was generally perceived as a threat by these farmers, most Grand River dairy farms are highly self-sufficient and viable, with very few having to resort to off-farm income. This is also a sign of the relative viability of dairying relative to most types of farming in Ontario, where off-farm income often accounts for about half of farmers' income.

Stress was positively correlated with respondent's education ($r = .255$; $p \leq 000$), but showed a negative correlation with respondent's perceived quality of life in 1997 ($r = -.305$; $p \leq .000$). Concern about the levels of milk

Table 9.6

Perceived quality of life of Grand River Watershed dairy farmers compared with others groups

Other groups	Mean response[a] (0 to 4)	% indicating much better[b]
US dairy farmers	2.77	22.9
Urban Ontarians	2.56	15.8
Canadians in general	2.52	4.5
Other Ontario farmers	2.52	4.7
Farmers of their parents' generation	2.45	9.8
People in their local town	2.40	8.1
Other Ontario dairy farmers	2.15	1.2
Nonfarm residents in their area	2.08	4.1

a Responses ranged from "somewhat worse" (0) to "much better" (4).
b Percentage of respondents whose answer was "much better" (4).

prices tended to be highest among those with high school education (see Table 9.7). The higher the farmers' educational level, the more stress they experienced from the amount of work they had to do. Generally, those farmers with high school education found their labour costs to be the most stressful. Production costs were also usually of most concern to those with high school education. The younger the farmer, the more stressed they tended to be by their animals' health. The amount of work was least stressful to the older farmers (over fifty-five years of age). Production costs and labour costs were usually most stressful for middle-aged farmers.

Farmers' own health was more stressful for those operating farms larger than 200 acres ($F = 3.19$; $p \leq .025$). Farmers found their amount of work the most stressful on farms in the range of 151 to 200 acres ($F = 4.77$; $p \leq .003$), while labour costs were seen as least stressful on farms of this size ($F = 2.70$; $p \leq .047$).

The cross-tabulation between age and herd size (Table 9.8) points to interesting motivational differences. Responses from the middle-aged group of farmers suggested the greatest divergence of goals. Herd expansion was a high priority for some, while others contemplated leaving the industry if marketing conditions changed. Rather than having a "disappearing middle," Ontario is more likely to experience a "strengthening middle" as herds consolidate within this group but are kept from ultimate concentration by the overall number of participants who intend to remain active milk producers. This group, however, by virtue of its high cost of production (probably resulting from the ongoing consolidation process), will remain the most vulnerable to price variations for several years.

Our survey responses indicate that while profit is a strong necessity, it is not sufficient by itself to create the high quality of life currently perceived

Table 9.7

Analysis of variance of significant stress factors with respect to educational level of Grand River Watershed dairy farmers

		Educational level			Analysis of variance	
Stress affecting life		Primary or less	High-school	Post-secondary	*F*	*p*
Debt	Mean	1.15	1.78	1.54	5.30	.006
	SD	1.05	1.24	1.11		
	N	67	69	52		
Weather	Mean	1.34	2.27	1.98	13.83	.000
	SD	1.05	1.10	0.96		
	N	65	70	52		
Environmental quality	Mean	0.66	1.06	0.77	4.08	.019
	SD	1.05	1.10	0.96		
	N	64	66	52		
Own health	Mean	0.71	0.94	1.12	3.10	.048
	SD	0.72	0.84	1.11		
	N	65	68	52		
Milk price level	Mean	0.43	0.80	0.48	3.34	.038
	SD	0.81	1.04	0.75		
	N	65	66	52		
Amount of work	Mean	1.24	1.81	1.96	6.69	.02
	SD	1.12	1.14	1.25		
	N	67	68	52		
Labour costs	Mean	0.84	1.41	1.10	3.91	.022
	SD	1.08	1.17	1.19		
	N	61	63	49		
Production costs	Mean	1.24	1.96	1.77	7.57	.001
	SD	0.90	1.23	1.13		
	N	66	68	52		
Government cutbacks	Mean	0.56	1.37	1.13	11.23	.000
	SD	0.96	1.04	0.99		
	N	64	68	52		

by Ontario's dairy farm operators. We looked even closer at the survey responses to attempt to discern any other emergent pattern that might help us predict the direction of change. Interpreted against the changing demographic profile of the dairy farming population, we might be able to recognize a pattern more clearly if the differences between farms owned by various age groups are linked with differences in attitudes.

Table 9.8

Farm numbers in the Grand River Watershed by age of farmers and herd size count

	Herd size category							
Age of farmers	1 1-19	2 20-39	3 40-59	4 60-79	5 80-99	6 100-119	7 160-179	Total
20+		5	2	1				8
30+		19	15	3		1		38
40+	5	24	22	11	2	1		65
50+	2	19	14	5	2	3	1	46
60+	2	9	6	2	1	1		21
70+		4	4	3				11
80+				1				1
Total	9	80	63	26	5	6	1	190

Mostly, the large herds in Ontario belong to farmers at either end of the demographic profile, namely, under thirty years or over fifty years. The older group expresses belief in the need to have a large herd both to be cost-efficient and to achieve a satisfactory income level. The younger owners of large herds express less desire for expansion than their middle-aged, less efficient counterparts. Also, they are not likely to be in as good a financial position to expand as their older contemporaries. Any efficiency gained by having a larger herd will likely be eaten up by the financial costs of acquiring it. Thus, it seems likely that herd dispersals, which occur as farmers retire, will transfer stock into the hands of less efficient producers near the centre of the age profile and near the centre of the herd size profile. It is in this group where farmers more strongly believe that greater profitability can be achieved only with larger herds. While it is not likely that milk output per cow will change much with these transfers, the average cost of milk production in Ontario will likely not decrease. Because of this, the overall competitiveness of the Ontario industry may not improve for several years as the total number of animals involved in this process is nearly 60% of the total provincial milking herd.

If our expectations prove correct, extreme pressure may be placed in the future on marketing managers to develop a clear understanding of the true cost of producing a hectolitre of milk in Ontario compared with other regions. Should any prolonged reductions in milk prices occur, Ontario's middle-aged dairy farmers may be forced out of business, leaving a cadre of young, cost-efficient, income-satisfied producers to carry on, confident in their ability to compete in an open market.

The changes suggested by the data may take decades to unfold. Thus, a return to chaotic times may lie ahead for both farmers and policymakers.

Having a group of young, confident producers arrive just as marketing conditions change may be a godsend for the industry. On the other hand, the pressures arising from the adjustment process in the middle of the age and herd size profiles could not have come at a worse time. That this process may be chaotic for the next several years is partly the result of decades of stability and slow change among producers.

As these changes that spell possible new directions for dairy farming in Ontario are occurring, other levels in the industry are feeling the effects of recent events that may dramatically alter business conditions throughout the dairy sector.

At the policy level, Canada is a participant in the WTO, which was recently born out of GATT. Participant countries have pledged support for the gradual removal of trade barriers and market-distorting subsidy practices within their borders. In heretofore highly regulated sectors of agriculture with significant market interventions, the new WTO rules will challenge both structure and economic potentials.

Policy changes occurring at the highest levels of the industry will exert economic effects at the farm level. The way dairy farmers react to these changes and the managerial orientation of younger milk producers will play a key role in defining the future of dairying and will affect other participants in the dairy product supply chain.

Younger farmers (under forty-five years) expressed higher than average confidence about their ability to adjust if milk marketing conditions were to change. They did express the opinion that the present marketing system was contributing positively to their quality of life. Tacitly, they assumed that milk prices would fall without the current marketing system. However, this age group exhibited the greatest frequency of over-quota shipments at the world price and also the greatest admission of this being deliberate. Almost half said they were regularly shipping over-quota milk.

This suggests competitive confidence, income satisfaction, and managerial acumen all at once. They were relatively unconcerned about free trade issues. Their age group expressed the greatest degree of positivity towards NAFTA and trade issues. More farmers reported experiencing a positive effect from the signing of GATT (37.6%)[3] than from the signing of NAFTA (23.3%). Interestingly, farmers with larger acreages tended to view NAFTA as being detrimental to their quality of life, while at the same time reporting a positive effect from the signing of GATT.

The majority of those with a high school education or above saw recombinant bovine somatotropin (rbST) as detrimental, while most of those with less education said they did not know what the consequences would be. As

3 Although the WTO had been created at the time of the survey, respondents were more familiar with of the signing of GATT, and that is how the question was asked.

well, over half (51.1%) expected a positive effect to result from having a Canadian food self-sufficiency plan, although 30.8% stated that they "did not know" what the effect of this action would be.

Middle-aged dairy farmers showed no discernible consensus on free trade. However, this group displayed the greatest negativity towards NAFTA. Because of their numbers, middle-aged dairy farmers currently carry the greatest weight in the politics of the industry. All of the dairy farmers reported sensitivity to milk price changes. One-fifth (19.8%) said that they would leave dairy farming if there was a price reduction of 4%. One-quarter (25.7%) said that they would discontinue if the reduction was between 5% and 9%. One-third (34.7%) said that they would stay in the dairy business unless the milk price reduction was greater than 15%. Concern about the levels of milk prices and production costs tended to be highest among those with high school education.

The differences in outlook between these two age groups suggest several things about Ontario dairying in the future. Blend prices for those liberally shipping over-quota are likely higher than would be the case without production quotas. These farmers will probably align themselves with the centre in resisting change. Also, they likely know their competitive position and will continue to ship milk at the world price, perhaps inadvertently making the point to the WTO that our industry is free-trade friendly.

On the other hand, their confidence may spark a search for ways to rid themselves of a significant costly barrier to expansion, namely, production quotas. As long as quotas remain in place, this group will most likely pursue business as usual. However, by overtly not seeking expansion on their farms, they will also not contribute toward quotas prices holding high value on the exchange. Significant changes in the industry may simply have to wait until the competitive struggle consolidating herds among those in the centre has played out. This process will likely last longer than ten years, as the average age of the participants is only around forty-five years. By then, the now younger dairy farmers will have polished their skills and improved their financial position while waiting for the opportunity to put an end to quotas (and their costs) and leapfrog the 8,000 litres per cow per year (MacDonald 1999, 21) barrier in one quick push.

Compared with many of the herds producing 12,000 litres per cow per year in New York state and elsewhere, Ontario's average productivity is relatively low. However, some dairy farmers at the average level of productivity in Ontario may simply have spent less to acquire their herds (that is, they have low-cost cows) and are economically efficient. The Ontario Dairy Farm Accounting Project (ODFAP) data (ODFAP 1999) show farms of high profitability at various levels of per cow productivity. While our survey participants had average net incomes approaching $65,000, the ODFAP data showed net incomes of $55,000 for the province as a whole.

Discussion and Conclusion

Viability can be measured by farmers' net farm income and their degree of dependence on off-farm employment. With a sample of southwestern Ontario farmers overlapping the Grand River Watershed, McCoy and Filson (1996) discovered that farmers' perceived quality of life tends to be inversely proportional to the amount of work they had to do off the farm in order to remain economically viable. The present survey found, however, that off-farm work contributes a very small part of Grand River dairy farmers' total net income – a sign that most Grand River dairy farms are highly self-sufficient and viable, especially for the male farmers, whose resort to off-farm income was considerably less than that of their wives. This is also a sign of the relative viability of dairying relative to most types of farming in Ontario, where off-farm income often accounts for about half of farmers' income. Thus, at least in the Grand River Watershed, we have a farming system with viability but somewhat less than optimal productivity.

On the other hand, the Ontario dairy industry has breeding stock that is on par with the best in the US and the Netherlands (that is, the world), notwithstanding the fact that these cows are not being deployed to produce the 12,000 litres per cow that is being achieved elsewhere. On the other hand, these Grand River cows have a less stressful life than many higher-producing American cows, in part because they are milked less often and do not use rbST. One may still ask whether there is sufficient incentive to be productive under supply management. Given the fact that the quota level is tied to the total amount of milk one is allowed to deliver and not the number of cows in one's herd (Dairy Farmers of Ontario 1998), there is definitely some incentive to be efficient.

Factors other than higher profit must come into play when a dairy farmer decides to remain at a certain size. Attitudes are tied to immediate economic circumstances and need, but history and experience play important roles in shaping a farmer's managerial outlook and approach. Because the demographic profile of the population of Ontario dairy farmers is entering a period of rapid change, the future of the industry will be determined partly by the numbers. By this we mean that farmers occupying different demographic divisions (age groups) and having different attitudinal characteristics, especially where ages differ by a generation (twenty or more years), will collectively influence the direction of the industry in a manner suggestive of a game having only a few players.

The most marked differences between age groups were in those aspects having to do with the future – theirs and the industry's. Older farmers with a long involvement in the industry (>50 years old with >25 years in dairying) expressed concern over the possibility that the WTO or NAFTA would spell the end of the current institutional structure of dairy farming.

They worry about production quotas (currently a highly valued business asset) disappearing before they have a chance to realize on their capital value (sell them). They typically view their quota holdings as a significant part of their pension. They rarely admit to the realization that it is precisely those quotas that have yielded the favourable milk prices that enabled them to build a large part of their non-quota capital value. These farmers are more likely to view trade liberalization as a threat. They may also be more prone to quick decisions about leaving the industry.

Supply management seems to be regarded as either a blessing or a curse, depending on one's perspective. Many economists and dairy farmers think that supply management results in too many cows producing too little milk, and therefore that it results in inefficient resource allocation. Orderly marketing enables the individual to embrace a range of production efficiencies, thereby taking the edge off the dictum of competitiveness, which may act as a damper on the drive towards participation in the global market.

Felstehausen (1993) found, for Wisconsin and Mexico, that there was no necessary link between high productivity per cow and high efficiency because the real question was how much is spent per cow to get there. Modern technology makes it possible for dairy farms to be efficient and profitable at all sizes from small to large herds. Therefore, he concludes, there is no underlying economic force driving the industry towards large herds. In the long run, sustainability may occur at a lower level of productivity.

Others question how heavily large cow herds with high productivity tread on the environment. They are dependent in the US on rbST, as well as on highly concentrated feeds produced from crops in turn dependent on fertilizer and pesticides. They suggest that dependence on "high-octane" inputs may not be sustainable, just as the Green Revolution was not sustainable throughout most of the Third World (Redclift 1987). Still, the Green Revolution philosophy persists in North America, implying that an important lesson has not yet been learned.

Some producers are driven to derive more money ("maximizers"), while others are satisfied with a reasonable income ("satisficers" in Professor Simon's words). Times of prolonged low prices may force everyone to be a profit maximizer. Conversely, in times of high prices, some producers whose herds are less efficient (and are "satisficers") will not push as hard towards more production efficiency (Simon 1957).

Thanks to supply management, both types of managers continue to operate in Ontario. Unregulated free enterprise can result in everyone being forced to adopt the habits of a profit maximizer or fail. From one point of view, such a system is more likely to be a treadmill similar to Cochrane's concept (Cochrane 1958). A sustainable system may be one that allows people to choose their management style.

Long-held opinions about appropriate practices in dairy farming centred on profitability as the main criterion of success. However, a careful look at the survey responses reveals that there are other motivations besides profit that hold dairy farmers' attention. Obviously, significant amounts of profit are being earned and individual farm incomes vary across a wide range. This range of earned income corresponds to a range in herd sizes. Generally, larger herds generate larger incomes. However, income per cow varies greatly even within groups of farms with similar herd size. Why? Apparently, a similar process is at work in Ontario as in Wisconsin, where farms of similar efficiency (net income per cow) could be found at all sizes (Felstehausen 1993).

We conclude, therefore, that whereas this supply managed dairy farming system is not the most productive of dairy farming systems, it is a very viable system and one that provides room for a wider range of managerial styles.

10
Water Quality Initiatives in the Crowfoot Creek Watershed, Alberta

Georgina Knitel and Alfons Weersink

Chapters 7 and 8 examined the potential abatement costs associated with alternative manure management systems. In both cases, models were used to predict the levels of net farm income and residuals from livestock production. The models could be simulated to assess *ex ante* the levels of these sustainability indicators for hypothetical farms. This chapter examines the *ex post* relationship between the environment and farm income for actual farmers taking a proactive approach to water quality improvement.

In Alberta a unique grassroots group of agricultural producers and neighbouring residents, the Crowfoot Creek Watershed Group, has begun to involve its community in planning and implementing activities that have a positive impact on the Crowfoot Creek Watershed. Community leaders, in partnership with other stakeholders, have initiated efforts to research and demonstrate BMPs and educate the community about them to improve the quality of water within their watershed through activities that are restoring and protecting riparian areas.[1] While the environmental impact of these efforts is positive, the on-farm economic benefits of implementing the BMPs do not always offset the implementation, maintenance, and opportunity costs.

The purpose of this chapter is to determine the on-farm income effects and the associated environmental improvement of four BMPs undertaken by the Crowfoot Creek Watershed Group. The producers within the Crowfoot Creek Watershed are becoming more aware of the negative impact that their own practices can have on water quality. They are also becoming more aware of the potential benefits that various BMPs can have on their watershed. However, there is little information available to these producers on the specific cost-effectiveness for each BMP. This economic information

1 Riparian buffers provide multiple benefits in the form of ecological economic goods and services such as recreation, preservation of biodiversity, and water purification (Qui and Prato 1998).

would assist interested producers in assessing the financial implications of implementing a given BMP and assist regulators in determining the level of compensation necessary to encourage producers to adopt BMPs.

We begin by discussing the Crowfoot Creek Watershed and recent studies assessing the factors affecting water quality. The third section reviews the results of a survey of producer awareness and attitudes towards water quality. The fourth section determines the on-farm profitability of four BMPs that have been undertaken in the Crowfoot Creek Watershed in an effort to enhance water quality. The concluding section discusses the policy implications of the results.

The Crowfoot Creek Watershed

The Crowfoot Creek Watershed is located 85 km east of the city of Calgary, Alberta, in Wheatland County. It encompasses approximately 1,600 km^2 in an area of intense agriculture, irrigation, and oil and gas activity. Crowfoot Creek is a small tributary of the Bow River fed by the natural flows of runoff from snowmelt, irrigation, and high rains occurring in the basin. The natural flow of Crowfoot Creek is augmented from May through October by deliveries and/or spills of irrigation water from the Western Irrigation District. Water within the creek is used for domestic purposes, irrigation, livestock watering, and creation of wildlife habitat through constructed wetlands. Excess water from the drainage is discharged into the Bow River. The Bow River is a sub-basin of the South Saskatchewan River that flows into southern Saskatchewan, which in turn is a sub-basin of the Lake Winnipeg Basin that drains into Hudson Bay. Hence, the activity within the Crowfoot Creek Watershed has the potential to impact a large geographical area.

Crowfoot Creek Watershed Group

The Crowfoot Creek Watershed Group (CCWG) was formed in January 1999 as a proactive grassroots group of concerned community members who gathered in response to a series of reports regarding the quality of water in their watershed. The first of these reports was released in 1995 after the Wetland County Agriculture Services Board initiated a Surface Water Quality Data Assessment Project. The results from this project led a number of concerned residents within the watershed, supported by Alberta Agriculture officials, to initiate a sampling program that would further define the challenges within the watershed. The program, called the 1996-1999 Crowfoot Creek Agricultural Impacts on Water Quality Study, identified several areas of concern, including levels of fecal coliform bacteria, pesticides, and phosphorus and nitrogen nutrients that at times exceeded the Alberta and Canadian Water Quality Guidelines. Since the watershed's primary activity is agriculture, it appeared that agricultural practices were contributing to the declining water quality.

The CCWG is made up of approximately sixty-five members, either agricultural producers or residents living within the Crowfoot Creek Watershed. The group is producer-driven, and has taken responsibility for water quality within their watershed. The CCWG's vision is to have a healthy and sustainable watershed. Its mission is to work with community producers and residents towards the adoption of practices that will result in water leaving the watershed in a condition equal to or better than when it arrived (*Prairie Water News*, 1996).

The CCWG intends to achieve its vision through research and long-term monitoring, education and awareness programs, and demonstration projects. Various means of communication are used, such as newsletters and other written communications, media, meetings, and demonstration tours. The group works with local, provincial, and federal stakeholders towards increasing community awareness of water quality issues and watershed management in the Crowfoot Creek Watershed. The CCWG believes that the adoption of BMPs will promote a sustainable environment and subsequent sustainable economy for future generations.

The CCWG has linked up with several corporate and public organizations that have contributed to the facilitation of water quality projects within the Crowfoot Creek Watershed. Many of the partners have stated that they have been able to provide their support through funding primarily because the CCWG is a very organized and cohesive group that is working closely with the stakeholders within the watershed. Current partners of the CCWG include: Alberta Agriculture, Food and Rural Development (AAFRD), Wheatland County, Western Irrigation District (WID), Prairie Farm Rehabilitation Administration (PFRA), Ducks Unlimited, Pheasants Forever, Health Authority 5, Husky Oil, PanCanadian, Elite Swine, Village of Standard, and Town of Strathmore. Additional funding has been provided through the Alberta Environmentally Sustainable Agriculture (AESA) program and recently from the Community Riparian Program (CRP).

The Crowfoot Creek Watershed Study

The Matching Grant Program of the Alberta Agricultural Research Institute funded the Crowfoot Creek Watershed study, with additional contributions from various partners in the study. These partners included stakeholders and organizations affiliated with them. The study was conducted over a period of four years from 1996 to 1999. The objectives were to determine practices that were contributing to the deterioration of water quality in the Crowfoot Creek Watershed, and to identify the land uses having such effects.

The Crowfoot Creek Watershed is a natural prairie watershed but is influenced by irrigation projects. Most prairie watersheds experience water flow predominantly during the spring snowmelt period only. Through irrigation, however, additional water is brought into Crowfoot Creek that normally would not enter the watershed. The level of flow for Crowfoot Creek is therefore higher than would otherwise be the case in most prairie watersheds during the period after the snowmelt until the end of the irrigation season. This enhanced flow from irrigation return has a direct effect on the concentration and mass load of parameters within Crowfoot Creek.

Monitoring equipment was installed at twenty-eight sites within the Crowfoot Creek Watershed. The sites were selected by the producers within the basin and other partners in the project, considering the land use, hydrology, and biology of the immediate area. The most common land uses were annual crops and summerfallow, feedlots, native and improved pasture, and perennial forages. Seventeen sites were located on Crowfoot Creek or its tributaries. Six sites were located where there were high flows of water originating from the Western Irrigation District returning into Crowfoot Creek. Four sites were located where water was flowing into the watershed through the WID infrastructure.

During the four years studied, most water flow through the watershed was from irrigation return flows; in 1997 there was uncharacteristically heavy spring runoff resulting from the melt of heavy snowpack. Annually, flow heightened with increases in use of irrigation as well as with sharp increases during and immediately after rainfall events. The sites were sampled weekly from May through October over the four years of the study. Data were collected on flow volume, water depth, and flow velocity. During snowmelt or various rainfall spells, the samples were collected daily.

The Crowfoot Creek water quality research confirmed much of what the original studies had concluded: the subject water basin was contaminated; some of the contaminants were brought into the water basin from sources prior to flow entering the basin, but most of the pollutant sources were from within the basin. Once the water was within the watershed, its quality began to decrease, specifically in the months of March through October of each year. The quality of the water within the Crowfoot Creek Watershed was often considerably inferior to the Alberta Water Quality Guidelines.

Snowmelt runoff caused contamination of the water within the watershed. The runoff, combined with plant breakdown in the spring, resulted in a high concentration of nutrients in the water. Both nutrients and fecal coliform bacteria registered higher levels in the spring and slowly decreased through the summer. Once the cattle were brought out to the grassland, however, the fecal coliform count increased. It dropped again once the cattle were off the land. The fecal coliform counts often exceeded the Alberta Water Quality Guidelines for irrigation, contact recreation, and potable water.

Each occurrence of rainfall resulted in a temporary increase in contaminants in the water with erosion runoff, but phosphorus remained at higher levels throughout the rest of the year. The count of contaminant levels increased during the summer months while the cattle grazed near the creek, and then dropped off as the cattle were moved to winter pasture. Sediment in the water was higher from soil erosion caused by rainstorms. Wetter soil caused enhanced runoff, so there were higher levels of nutrients in the runoff during rain events. Often, if crops were not mature enough to cover the land and bind soil, high particulate form dominated in the runoff. It appeared that there was less particulate matter in the runoff in areas of the watershed that had more grassland.

Cattle also contaminated the water, primarily by directly accessing the stream. Fecal coliform bacteria levels were correlated with the grazing patterns of the Crowfoot Creek Watershed. Higher fecal coliform counts were found in sub-basins with grassland adjacent to Crowfoot Creek and other waterways and where cattle had open access to the water. The count increased during the summer months while the cattle grazed near the creek, and then dropped off as the cattle were moved to winter pasture.

Regarding pesticides, samples were analyzed for the presence of a total of twenty-five different compounds; fifteen were detected. Of these, dicamba, MCPA, MCPP, 2,4-D, and atrazine were detected consistently. Dicamba, MCPA, and 2,4-D break down naturally, but the research showed that they tended to increase throughout the year or remained constant. The sources of these compounds must therefore have been constant, and the pesticides moved through the stream. Dicamba is a primary pesticide used by the WID along its waterways. MCPP and atrazine tended to decrease as they worked through the watershed. It was therefore concluded that their source was upstream from the watershed. It was also noted that atrazine was not used by the producers within the Crowfoot Creek Watershed, so the contamination had to be from an upstream source, most probably the city of Calgary's sterilant program for areas where vegetation is not wanted.

The study has provided further insight into and awareness of the water quality issues faced by the Crowfoot Creek Watershed Group. Various stakeholders within the group have elected to implement better management practices to enhance the water quality within the watershed. The group has also undertaken initiatives to increase community awareness of water quality issues by sharing information and developing demonstration sites of better management practices. To enhance water quality within the Crowfoot Creek Watershed, however, all local industry, including agriculture, and residents within the watershed should implement best management practices. The Crowfoot Creek Watershed Group has recently conducted a survey of its stakeholders within the watershed to determine their current practices and levels of awareness of water quality issues. The survey should help determine the producers' attitudes towards implementing BMPs.

Producer Understanding of Water Quality

A consultative survey of 100 agricultural producers within the Crowfoot

Creek Watershed was conducted during the late summer and early fall of 2000. The purpose of the study was to raise awareness and understanding of water quality issues and opportunities in the CCWG. The information from the survey has been shared with the stakeholders. Members of the CCWG wanted to use the survey results to determine whether there were any obvious correlations between the original water-monitoring results and the barriers and opportunities for future agricultural practices. The issues surrounding the correlations could then be discussed and possible solutions suggested. The study also identified the stakeholders within the CCWG who would be willing to consider adopting suggested practices.

General Awareness

Of the producers surveyed, 95% were aware of the definition of a watershed and the fact that they resided in the Crowfoot Creek Watershed. However, only just over half (58%) of the producers were aware of the definition of best management practices. Some 61% selected water quality as very important and 37% rated it as important. A large majority (91%) agreed that everyone is ultimately responsible for preserving water quality in the watershed for future generations. Sixty percent of the producers surveyed joined as active members of the CCWG.

Current Livestock Practices

The survey indicated that a significant number of producers (77%) within the Crowfoot Creek Watershed did not use alternative watering systems and did not limit livestock access to water bodies. Half (50%) of the producers used the creek as a water source for livestock from spring to fall, and 20% utilized the creek in the winter months.

Crop and Hayland Management

Most producers were practising conservation tillage, of which 38% practised direct seeding, and 32% of all producers practised minimum tillage. Many (63%) of the crop producers utilized soil testing before applying elemental fertilizers. At the time of the study 13% of the total cropland, at 10% of the total land, was summerfallowed. Some 8% of the producers had established field shelterbelts. All producers attempted to keep stubble up to reduce runoff erosion, and 80% were using various best management practices to minimize impacts on water quality from their spraying methods, such as maintaining the recommended distances away from watercourses when spraying.

Manure Management

Of the farmers within the Crowfoot Creek Watershed who utilized manure as a source of fertilizer, application rates were often based on historical practices (60%) or visual crop response (40%). Most producers (65%) applied the manure every six to twelve months, and applied it on cropland (85%) where it could be easily incorporated. Over 60% incorporated manure into the soil, and 80% agreed that it should be incorporated immediately after application when possible. All producers agreed that manure should not be applied in the winter, but they were not asked whether they practised their belief. Most producers did not test soil (92%) or manure (98%) for fertility before applying manure to the land. Most producers (77%) were convinced that it was not necessary to test manure for nutrient levels before spreading it on the land, perhaps due to the belief that all manure holds relatively the same nutrient level.

Riparian Management

Approximately half of the producers (51%) did not maintain or make attempts to enhance the crop and pastureland that interacted directly with the water and soil. Many producers maintained grass waterways (80%), buffer strips (70%), and riparian areas (48%).

Challenges and Concerns

Of the 100 Crowfoot Creek Watershed producers surveyed, nearly half (48%) had visually observed that water quality was worsening in the Crowfoot Creek Watershed and the linked waterways. Observations included a changed colour and debris in the water. Some 32% said that the water quality appeared unchanged. Most producers had environmental concerns regarding four themes: 48% were concerned with livestock, 43% with agricultural chemicals, 20% with urban contamination, and 15% with impacts from the oil and gas industry. Other issues cited included irrigation impacts, soil erosion, manure management, and drinking water.

The perceived cost of utilizing best management practices was cited as the most common challenge (51%) for producers to take action to improve their methodology and subsequent impact on the Crowfoot Creek Watershed. The cost of fencing (44%) and access to water in all pastures (37%) were the most common challenges that limited the use of rotational, time-controlled, rest-rotation, or landscape grazing for livestock producers. Most of the surveyed CCWG producers did not compost manure because of the perceived high cost (34%) or because they believed that a composting system would not be practical for their operation (61%).

Opportunities for the Future

There were many suggestions for an action plan that could enhance water quality within the Crowfoot Creek Watershed. Most producers cited demonstrations (48%) and newsletters (51%) as a preferred method of learning compared with site-specific problem solving (34%) and one-on-one consultations (33%). Other suggestions included:

- utilizing educational tools and information from other communities and similar organizations
- continued partnerships and new partnerships that facilitate the liaising with local and regional organizations such as the Cows and Fish Program, Autobahn Society, Alberta Reduced Tillage Initiative, and so on
- correlating the survey data with water quality monitoring results to help identify geographic areas and practises to improve upon
- education for the next generations through school programs, clubs, and media
- focus on understanding and improving water quality and quality relationships in the watershed through long-term monitoring and improved partnering with the irrigation district.

The Crowfoot Creek Watershed study revealed that water quality often deteriorated from the inflow into the water basin to the outflow. The results of the study point to agriculture, the primary industry within the watershed, as a contributor to pollution in the Crowfoot Creek. The producer survey determined that of the 100 participants surveyed, most were cognizant of water quality issues and various practices that affected water quality within their watershed. The participants also demonstrated a desire to make the effort to maintain a healthier watershed by educating themselves and implementing best management practices. However, what is the cost-effectiveness of implementing such practices, and who should pay?

On-Farm Costs and Benefits of Best Management Practices

Within the past two years, agricultural producers in the Crowfoot Creek Watershed Group have taken the various sources of information and implemented best management practices at their operations on a voluntary basis. These BMPs will reduce the contamination to the watershed; however, follow-up monitoring has been underway only recently. In time, the data from this monitoring effort will be compared with the baseline measures established in the four-year database, and the success of the BMPs can be quantified. Such comparison will not be without difficulties, however, as it is difficult to measure the BMPs' direct benefits to the watershed, since most contamination from the agricultural operations within the study area is nonpoint source pollution. In addition, water quality is highly variable due to dynamic annual weather conditions and responsive changes in management practices.

Four individual BMPs will be reviewed below, each one at a different location within the Crowfoot Creek Watershed. Accurate estimates of the water pollution abatement costs are critical in the producers' assessment of whether or not to adopt BMPs. Water pollution abatement costs to be calculated are the direct costs (installation costs, maintenance costs, and opportunity costs) to the producer. Installation costs such as capital expenditures, seed, and equipment costs are incurred when the BMP is established. Maintenance costs include operating costs necessary to support the BMP, such as land and labour. The opportunity costs are measured by the forgone net return on the land used in the BMP. Indirect abatement costs, such as water pollution control costs embedded in goods and services, will not be calculated. The direct economic benefits to the producer will be any financial return resulting from the BMP.

Farmer 1

Farmer 1 has pastureland next to the creek. On one fifty-three-hectare parcel of pasture, the only access the cattle had to water was the creek located on the southwest corner. The riparian area was suffering from trampling and the water quality was affected by the presence of cattle in the creek.

Farmer 1 elected to provide an alternative watering source that diverted the cattle away from the creek. He did this for environmental as well as economic sustainability reasons. Farmer 1 was aware of the Crowfoot Creek sample results, and he shared in the concerns regarding the environmental and economic sustainability of his farming and ranching operation. Initially, the primary goal was to limit cattle access to the stream, thereby reducing the contamination to the stream and rehabilitating the riparian area. Farmer 1 noted that, given a choice, the cattle preferred to drink from the trough rather than climb the banks of the stream and wallow in the muddied water, from which it had to drink. A direct benefit to Farmer 1 is that a cleaner water supply is known to improve the rate of weight gain in cattle.

While Farmer 1 was considering limiting access to the stream by providing an alternative water source, he realized that a significant economic benefit could accompany the environmental advantages. By relocating the water source to a central location, the pastureland could be more efficiently utilized, allowing for greater carrying capacity on the fifty-three hectares. Farmer 1 fenced off four quadrants in the fifty-four-hectare parcel, with the 500-gallon water trough at the meeting point of the quadrants. Subdivision of the pasture into smaller quadrants allows for rotational grazing. The cattle distribution over the grazing season improved, since the animals are forced to graze the upland forage and thereby utilize the overall range landscape more evenly. Shorter grazing periods enables the grass to rest and regain vigour when cattle are moved into the other pasture quadrants. Farmer 1 can also avoid grazing the land closest to the stream during the riparian area's most vulnerable times, such as during runoff, when banks are most fragile or when cattle effluent can most easily be carried into the creek.

In Farmer 1's case, there was a one-time installment cost of approximately $6,210 (Table 10.1). This cost included the capital investment in pumping and water storage equipment, fencing supplies, and the installation labour. The annual maintenance cost is approximated at $240, consisting primarily of labour to maintain the fence and service the windmill, pump, and water storage. The opportunity cost of the windmill and pump is estimated to be nil, since no significant portion of land was taken out of use.

The cost of the windmill pump watering system of Farmer 1 was shared by the CCWG, AESA/AAFRD, PFRA, Ducks Unlimited, Husky Energy, and PanCanadian Petroleum. The Crowfoot Creek Watershed Group did the planning and ordering of supplies, equipment, and resources for the setup of the pump, pipe, and dirt work. Farmer 1 did the ordering and purchasing of supplies, equipment, and resources for the tank, trough, and fencing. AAFRD provided technical support for site surveying, advice on the pump setup,

Table 10.1

On-farm costs and benefits of windmill pump watering system for Farmer 1

	Year 1 costs	Benefits in subsequent years
Implementation activities		
Windmill watering pump	$2,610	Improved rate of gain due to cleaner water source (value unknown)
Water tank	$500	Improved utilization of 53 ha: $1,920
500-gallon water trough	$300	Pasture (20 head × $24 × 4 months)
1,300-foot 2-inch hose/pipe/fittings	$1,350	
1-mile fencing with four gates	$500	
Delivery and assembly/adjustments	$300	
Installation labour	$650	
Maintenance activity		
12 hours labour	$0	
Opportunity costs		
Nil	$0	
Total annualized costs and benefits	$6,210	$1,920

and a portion of the funds for the pump costs through the AESA program. The PFRA provided technical support for the site surveying and pump setup. Ducks Unlimited, Husky Energy, and PanCanadian Petroleum provided funds to defray pump costs.

Farmer 1's windmill pump water BMP resulted in some economic benefits in addition to the environmental benefits to the watershed. Based on the assumption that the rent for grassland remains at $24 per cow-calf pair, which is the average rate currently being charged in the area as quoted by the AAFRD, the economic benefits have been estimated at a net present annualized value of $1,920. If Farmer 1 were to absorb the total cost of the BMP, the internal rate of return on the $6,210 total cost would be 17% after five years.

Farmer 2

Farmer 2 had livestock yard and corrals situated within fifty feet of the creek. The proximity to the waterway caused erosion and transport of manure to the creek, particularly during spring melt and rain events; runoff contaminated the creek.

Farmer 2 elected to build a simple drainage diversion system away from a cattle overwintering site. He constructed an earth burm downslope of his livestock corrals to divert runoff out to where it waters and fertilizes his pasture, and away from the creek. This was intended to help control the bacterial, phosphorus, and nitrogen loads in the runoff. Farmer 2 stated that he elected to implement the diversion burm because of his social conscience. The CCWG coordinator visited his operation, informed him of the water sampling results and subsequent concerns, and suggested possible solutions. Farmer 2 decided to construct a burm that had a low implementation cost and no maintenance or opportunity costs.

The one-time installment cost of the corral diversion project was approximately $1,000, which included equipment rental and labour (Table 10.2). There is no annual maintenance cost. The opportunity cost for the earth burm is also estimated to be nil, since no significant portion of land was taken out of use. For Farmer 2, there appear to be only minor immediate economic benefits resulting from the fertilizing of a portion of his pastureland to accompany the social economic value from the environmental benefits to the watershed.

Partners contributing to the corral runoff diversion system were Farmer 2, the CCWG, AESA/AAFRD, PFRA, Ducks Unlimited, and Husky Energy. The CCWG provided the design and ordering of equipment and resources to construct the burm. Farmer 2 prepared the site for work by moving equipment and materials that were obstructing the construction of the burm. Farmer 2 agreed to seed the burm to grass if necessary, and maintain the project for a minimum of five years. AAFRD provided the technical support and a portion

Table 10.2

On-farm costs and benefits of corral diversion project for Farmer 2

		Year 1 costs
Implementation activities	1-day grader	$500
	1-day bobcat	$300
	2 operators	$200
Maintenance activity	Nil	$0
Opportunity costs	Nil	$0
Total costs		$1,000

of the funds for the construction of the burm through the AESA program. The PFRA provided technical support. Ducks Unlimited and Husky Energy each provided some funds for construction of the burm.

Farmer 3

A significant portion of Farmer 3's land includes the Crowfoot Creek and a floodplain. Farmer 3 traditionally cropped right up to the creek, often contributing to nutrient, chemical, and soil erosion entering the creek during runoff.

As a result of becoming more aware of the health of the Crowfoot Creek Watershed due to the water sampling study and after much reading and conversation regarding erosion, Farmer 3 understood that his current practices were transferring nutrients into the creek. He decided to alleviate some of damage to the riparian area by planting a grass buffer strip on nineteen hectares of floodplain along the stream bank. The grass would work like a filter to reduce the impact on the water from nutrients, chemicals, and soil erosion.

The Farmer 3 grass buffer strip incurred a one-time installment cost of approximately $3,110 (Table 10.3). This cost includes the operating costs of grass seed and seeding. The grass mix used consisted of Reed Canary, Timothy, Smooth Brome, Orchard Grass, and Clover. The annual maintenance cost is estimated to be nil, since the buffer strip is to be left natural. The opportunity cost of the strip is approximately $822 ($45 per hectare), which is calculated as the forgone net income from crop production on the 18.25-hectare strip (Table 10.3). In the past, Farmer 3 cropped the 18.25 hectare as he did the adjoining land; he had practised a five-year rotation of canola, malt barley, wheat, wheat, and then summerfallow for weed control purposes. Approximately one in ten years, the 18.25-hectare area experienced a damaging flood and no crop would be harvested.

Table 10.3

On-farm costs and benefits of grass buffer strip for Farmer 3

	Year 1	Subsequent years
Implementation activities		
Grass seed	$2,610	nil
Planting	$500	nil
Pheasants Forever reimbursement	$2,610	nil
Maintenance activity		
Nil	nil	nil
Opportunity cost (see Table 10.4)		
18.25 ha/$73.08 net income	$1,334	$1,384
Pheasants Forever payment (18.25 × $10)	$1,834	$183
Total annualized costs/benefits	$1,834	$1,151

This grass buffer strip project of Farmer 3 was in partnership with the CCWG and Pheasants Forever. Farmer 3 ordered the seed and provided all the equipment and labour for seeding the grass mix. Farmer 3 also agreed to maintain the site for a minimum of five years. The CCWG contributed planning and advice regarding the buffer strip, and profiled the site through signage and tours. Pheasants Forever provided funding for the grass seed.

The Farmer 3 grass buffer strip BMP did not result in economic benefits but did provide environmental benefits to the watershed. Shortly after Farmer 3 planted the 18.25-hectare buffer strip, the local watershed coordinator arranged for Pheasants Forever to enter into a contract with Farmer 3. Pheasants Forever paid for the grass seed ($2,610) and agreed to pay about $10 per hectare for every year that Farmer 3 left the grass untouched. Thus, with the financial assistance of Pheasants Forever, the returns are $1,833.71 in year one and $1,151.21 in the subsequent years (see Tables 10.3 and 10.4).

Farmer 4

The primary source of water for the cattle of Farmer 4 is the creek that cattle had had free access to when grazing. One grazing location showed significant distress from the cattle accessing the creek.

Farmer 4 wanted to help repair and rehabilitate the riparian area and water quality by stabilizing the creek banks that his cattle were trampling. He installed a livestock access ramp with gravel and goetextile that provided easy access for the cattle without deterioration of the creek banks and muddying of the creek. He then fenced off the waterway to exclude all access to the creek except for the ramp, where the cattle would still contaminate the creek with excrement.

Table 10.4

Production costs and returns for Farmer 3 ($/acre)

	Canola	Malt barley	Wheat
Yield/acre	25	55	30
Expected market price	$5.85	$3.20	$4.50
Crop sales	$146.25	$176.00	$135.00
Direct expenses			
Seed	$12.00	$7.00	$7.50
Fertilizer			
Nitrogen	$12.00	$12.00	$24.00
Phosphorus	$7.60	$7.60	$9.50
Potassium	$0	$0	$0
Sulphur	$3.00	$0	$0
Chemical pre-seed	$0	$0	$4.50
In crop	$20.00	$20.00	$20.00
Pre-harvest	$0	$0	$0
Crop insurance	$5.46	$3.20	$3.95
Fuel, oil, and lubrication	$8.75	$8.75	$8.75
Machinery repairs	$8.00	$8.00	$8.75
Labour	$22.00	$22.00	$22.00
Summerfallow	$14.40	$14.40	$14.40
Total direct expenses	$113.21	$102.95	$123.35
Net income	$33.04	$73.05	$11.65
Average income for 5-year cycle of malt barley, canola, wheat × 3:	$28.66/acre ($73.08/ha)		

The one-time installment cost, including capital investment in the ramp supplies, equipment, fencing supplies, and installation labour was approximately $10,910 (Table 10.5). The annual maintenance cost is $480, which consists primarily of labour to maintain the ramp and fence. The opportunity cost to the windmill and pump is estimated to be nil, since no significant portion of land was taken out of use. The access ramp and fencing BMP did not result in any measurable economic benefit to Farmer 4, but there are likely to be social economic benefits that are not so easy to measure.

The watershed coordinator was able to facilitate the cost sharing of the access ramp. Partners that contributed to the access ramp and fencing system were Farmer 4, the CCWG, AESA/AAFRD, PFRA, Ducks Unlimited, Pheasants Forever, and the Western Irrigation District. Farmer 4 set up the fencing and provided two volunteers to assist with construction of the ramp. Farmer 4 also agreed to maintain the project for a minimum of five years. The CCWG did the planning, ordering, and purchasing of supplies, equipment,

Table 10.5

On-farm costs and benefits of access ramp and fencing for Farmer 4

	Annual costs	
	Year 1	Subsequent years
Implementation activities		
Permanent fence ($1.70/m × 961 m)	$1,634	
Extra wire	$245	
Posts	$245	
Double braces ($90 × 23)	$2,070	
Single braces ($50 × 6)	$300	
Plank fence	$240	
Steel gate	$120	
Rough-terrain charge	$120	
Geocell ($265/section × 12)	$3,180	
Crushed gravel ($11.60/cubic yard × 20)	$232	
Pitrrun rock	$500	
Labour	$960	
Shipping costs	$0	
Maintenance activity		
24 hours	$480	$480
Opportunity costs		
Nil	$0	$0
Total annualized costs	$10,910	$480

and resources to develop the access ramp, and paid for the materials used to set up the fence. AAFRD provided technical advice and assistance with the labour in the construction of the ramp. The PFRA provided technical support and a portion of the funding for the access ramp materials and construction costs. Ducks Unlimited and Pheasants Forever provided some funds for the fencing and materials. The Western Irrigation District provided technical advice and equipment for the access ramp construction, including one backhoe, one operator, and one labourer for one and a half days of dirt work.

BMP Conclusions

The cost effectiveness of each BMP varied significantly. The direct costs of the four BMPs in the Crowfoot Creek Watershed were highest for the access ramp and fencing of Farmer 4, followed by Farmer 1's windmill pump watering system and Farmer 3's grass buffer strip; the lowest direct costs were for the corral diversion of Farmer 2. Two of the four BMPs provided a positive direct return to the producers. The measurable financial benefits were highest for Farmer 1, with an internal rate of return of 17%, followed by

Farmer 3 with an internal rate of return of 10%. There were no direct financial benefits to Farmer 2 from the corral diversion BMP and to Farmer 4 from the access ramp.

The off-farm economic benefits from reducing water contamination are positive for all four BMPs. Farmer 4's access ramp reduces riparian area damage and direct contamination of the creek from cattle excrement. Farmer 2's corral diversion reduces contamination from cattle to the creek. Farmer 3's grass buffer strip enables the riparian area to be repaired and reduces contamination to the creek from pesticides and erosion. Farmer 1's windmill pump watering system enables the riparian area to be repaired and the pastureland to be improved due to better grazing patterns.

Considering the direct costs, direct financial benefits, and social economic benefits of each BMP, the windmill pump watering BMP of Farmer 1 is likely the most cost-effective, followed by the grass buffer strip of Farmer 3 and then the corral diversion BMP of Farmer 2. Although the access ramp and fencing of Farmer 4 provide benefits to the watercourse, their cost can be prohibitive.

Conclusion

The Crowfoot Creek Watershed Group is a grassroots proactive group of stakeholders who are primarily agricultural producers working towards a common goal of maintaining a healthier watershed. Their actions have had a direct impact on the Crowfoot Creek water basin and the water basins that evolve from it. The actions are resulting in social and economic benefits to many stakeholders within adjoining sub-basins, although a significant portion of the costs of implementing best management practices is still borne by the participating producer.

There are three identifiable reasons why the CCWG has been successful in having some of its producer members adopt BMPs. The first is group acceptance of responsibility for water contamination and water quality rather than blaming of individuals. It is relatively easy to measure the water quality of a watershed, and visible point source contamination can also be measured. It is difficult, however, to measure nonpoint source contamination that contributes to water pollution, and even more difficult to measure the improved water quality for BMPs at nonpoint source contamination areas. Understanding this, the CCWG has taken the stance that no finger-pointing would occur in the group, and that the group would collectively endeavour to implement BMPs that would enhance the water quality of the watershed as a whole. This understanding that it is the group's responsibility as a whole, rather than that of any specific individual, is a key to the success of the CCWG and the adoption of BMPs.

A second key to the success of the CCWG's voluntary adoption of BMPs is producer education. As producers become more aware of the facts regarding environmental damages and the benefits of various BMPs, some take a personal interest in and commitment to their environment. This responsibility towards their environment has resulted in the voluntary implementation of BMPs that will reduce contamination of the watercourse.

A third key may be the communication and organizational skills of the coordinator. The CCWG's coordinator was mentioned by each of the four producers studied in this chapter as a catalyst in the voluntary adoption of the BMPs. Through the efforts and follow-up of the coordinator, various producers in the CCWG implemented BMPs at a financial cost, without always getting a financial return.

The CCWG can serve as a model for future watershed groups wishing to enhance the environmental sustainability of agriculture and their watersheds. There are a variety of partnerships between individual producers, producer groups, government services, and other interest groups that are sharing in the costs of research, implementation, and maintenance of BMPs. Participation in most of these collective action programs is voluntary. An example is the Alberta Cows and Fish initiative, which has worked towards resolving riparian grazing concerns (Fitch and Adams 1998). Other examples of decentralized programs that subdivided the pollution control efforts by watershed rather than county or other political units is Wisconsin's legislated Wisconsin Non-Point Source Water Pollution Abatement Program (Castle 1993) and the Region of Waterloo's Rural Water Quality Program (Weersink et al. 2001). The decentralized approach is easier to sell politically in the farm community, but it may reduce the effectiveness of the program in environmental terms (Castle 1993). Yet, the collective action of a cost-sharing approach may result in the acceptance and implementation of BMPs that are cost-effective in enhancing water quality. If farmers are willing to incur at least some of the costs to improve water quality, and if the public is willing and able to supplement control costs through cost sharing, technical assistance programs, and tax incentives, there is considerable potential for improving environmental health (Philips and Veeman 1997).

11

The Ontario Environmental Farm Plan: A Whole-Farm System Approach to Participatory Environmental Management for Agriculture

John FitzGibbon, Ryan Plummer, and Robert Summers

In 1990, agriculture in Ontario came under increasing scrutiny with respect to its role as an agent of environmental degradation. Faced with the threat of increasingly rigorous regulatory control, the farm community responded with a movement to increase emphasis on stewardship and community (agricultural community) based action to deal with agri-environmental issues and problems. The Environmental Farm Plan (EFP) was the initial reaction of the farm community to the need for proactive environmental management in agriculture. It provided the basis for development of best practices that would lead to nutrient management planning, on farm source water protection planning, and the development of many local and national programs in which proactive compliance and stewardship, rather than command-and-control enforcement of regulations, provide the basis for environmental management. The EFP has been widely adopted in the farm community, especially in the animal sector and by those involved in intensive production who have seen greater concern expressed by the public with respect to agricultural practices.

The first step was the development of a vision for the agricultural environment. This was carried out through a series of meetings and workshops in which thirty-seven agricultural organizations sought consensus on defining the future for the environment in the agricultural area of Ontario. Using the vision statement and building upon it, the agricultural community developed an organizational structure called the Ontario Farm Environment Coalition (OFEC), which had the formal membership and structure of a general forum as well as a steering committee and various working groups to deal with specific issues. OFEC produced the Ontario Farm Environment Agenda, a policy document that set out specific issues and objectives, and strategies for achieving these objectives. A cornerstone of the agenda was development of the EFP, which was modelled on the "Farm a syst" program that had been developed at the University of Wisconsin. The

EFP was developed as a partnership between the farm community (OFEC), Agriculture and Agri-food Canada (AAFC), and the Ontario Ministry of Agriculture, Food and Rural Affairs (OMAFRA). It began as a pilot project aimed at 1,000 farmers and has gone through three subsequent project cycles to reach its present form. This action predated the Agenda 21 initiative. Local Agenda 21, the most significant output of the Rio Earth Summit, focused on local voluntary initiatives (Evans and Theobald 2003).

Early in 1993, OFEC established a nutrient management working group that responded to concerns that nutrients from agriculture were a serious source of water contamination. The group developed a model bylaw for use by municipalities that provided guidance for regulation of nutrients at the local level. As a result of pressure at the municipal level, there were constant changes to the bylaws and regulatory standards varied widely. OFEC then sought provincial legislation and consistent and predictable regulation. This was achieved in 2002 and the regulations were published in July 2003. The link to the EFP was established with the introduction of a short-form nutrient management plan (applied to farms with less than 150 nutrient units) (OMAF 2003b). This could be delivered through the EFP process, thereby eliminating the need for additional training and review.

The move to integrate aspects of nutrient management that are a regulatory requirement with the EFP, which is a voluntary program, is a first step in a move towards "soft governance" in the area of environmental management (Evans and Theobald 2003). It involves cooperation as the basis of compliance so that meeting the regulations is the result of being the "right thing to do" rather than "you must, or else." In the area of nonpoint source pollution, there has been a genuine lack of success with command-and-control approaches to regulation and compliance, and the move to a stewardship basis for compliance is now gaining wide acceptance (Deitz and Stern 2002). The EFP thus provides a model for a "soft governance" approach to environmental management based on community action.

EFP: The Approach

The approach taken in the EFP is a participatory planning and educational approach to planning and decision making. It follows the *precautionary principle* (Raffensperger and Tickner 1999) in that the actions taken under the plan respond to mitigation of risk of damage to the environment rather than the reactive approach towards much environmental regulation, which deal with insults to the environment after they have occurred. Under the precautionary principle, actions are taken on the basis of avoidance of damage that might reasonably be expected to occur, rather than being based on specific proof (Jordan and O'Riordan 1999). The EFP also follows the *principle of proportionality,* which stipulates that actions be taken in proportion to the risk of damage that might occur. This is dealt with in the plan through the use of a risk-rating system and development of an action plan, which deals with the priority risks.

In rating the risks, the EFP uses a set of best practices as norms and uses these to measure performance. The best practices *allow for margins of error (ecological space),* which makes the eventuality of unforeseen events less hazardous. The incorporation of considerations of human health, productivity of the environment, quality of the natural environment, and maintenance of resources integrates protection of the environment with protection of the resource base and protection of human and animal health. This integrated holistic approach as a system of interlinked elements provides for a real appreciation of *the farm as an ecosystem* and of protection of the environment as a component of a sustainable agricultural ecosystem.

The onus of action is on the farmer, who is the proponent of the agricultural activity. This places both responsibility and authority for environmental action on those who are undertaking the potentially risky activity. The linkage of responsibility and authority is essential to effective stewardship, which is at the heart of any precautionary management system. In its implementation, the farm plan also recognizes the role of the farmer as steward of the environment on behalf of society at large. This recognition is seen in the role of government in supporting the farm plan by providing financial support for both the educational component and the implementation of the plan. This societal responsibility has been exercised through financial support from the federal as well as provincial and local government agencies and from farm and nonfarm organizations.

EFP: The Process

The EFP process has a number of components. The first is awareness. Awareness is generated through the provision of information for both the general public and specific client groups. This has been achieved through the EFP implementation process carried out by the Ontario Soil and Crop Improvement Association and through both the producer group networks and the general agricultural media.

The second component of the EFP process is education. This is achieved through workshops that introduce the participants to the structure and process of assessment used in the farm plan. This is followed by discussion of the approach to rating the risks in the context of the farm operation and farming practices used. The participants then begin to undertake the step-by-step assessment of their farming operation by following the workbook sections that apply to the farming system of their enterprise. The initial assessments are done as a group activity with advice from the workshop coordinator, which allows for sharing of experiences and mutual learning in a supportive nonjudgmental situation. At the end of the day, the farmer returns home to complete the environmental assessment of the farming system. If issues come up, there is an opportunity to consult with the workshop coordinator and with other participants with whom a common bond has developed. At the same time, the results of the assessment remain confidential so that the assessment is not influenced by the possibility of some punitive response from a third party.

Upon completion of the risk assessment, the farm operator develops an action plan to deal with the risks identified. Priority actions are based partially on the magnitude of the risk and the judgment of the farm operator regarding the consequences that the exposure to the risk might have for the farm family and the environment.

The quality of the assessment receives a third-party review by the peer review committee. This is a committee of farm operators who are known for their expertise in farming and who have the confidence of both the framing community and the general public. The review process is required before the farmer can undertake implementation and be eligible for payment of the incentives provided for in the action plan. The peer review provides a measure of accountability for the decisions made in developing the action plan, and also provides a level of quality assurance for society.

Plan Implementation

At present, a total of 1,217 workshops have been delivered to a total of 25,178 participants. A total of 15,270 plans have been peer-reviewed and nearly 11,000 action plans have been implemented. This participation represents total of 8 million hectares of farmland covered by Environmental Farm Plans and 40% of farms in Ontario. The character of the participating farms is roughly representative of agriculture in Ontario (see Table 11.1). The types of action taken are for the most part protective of the productive capacity of the farmland (49%), followed by protection of water resources (32%), and farm health and safety (18%), with a small number of actions dealing with air quality (1%). Estimates of expenditures made by farmers range from $64 million to $104 million (in 2000 based on a 2% sample of farms that completed the planning process and claimed the incentive payment) (FitzGibbon et al. 2000a).

Table 11.1

Percentages of various farm types covered by Environmental Farm Plans in Ontario

Farm type	% under EFP
Cash crop	36
Dairy	15
Mixed farming	14

In addition, there has been a contribution of from 320,000 to 487,000 hours of farm labour to implement the action plans (FitzGibbon et al. 2000a). The mean expenditure per farm was $10,800 and the mean labour input by the farmer was 53 hours. On a county basis, participation ranged from a low of less than 10% to a high of nearly 70% of farms (see Figure 11.1). On 61% of the farms visited by the expert reviewers, the impact of the farm plan on risk mitigation was rated as strong to very strong. Of the farms included in the evaluative study, 53% of the environmental concerns that had been identified by the farms and confirmed by the peer review had been acted upon by the time the claim for the incentive support was filed. Factors that encouraged the farmers to take action included the awareness and knowledge that had been gained in doing an EFP. Over half of the farms reviewed in the study had revisited their plans and updated them, and 61% had undertaken environmental action above and beyond that identified in the plan. Finally, over 90% of the farmers indicated that they would voluntarily attend a one-day follow-up workshop (FitzGibbon et al. 2000a). The assessment of the impacts of farm plan implementation suggested that the program has been both successful and effective in improving farm practices that will reduce the risk of environmental degradation from agriculture.

Uptake of the program has been very good given the voluntary nature of the program. Indeed, adoption of the EFP has been increasing with time as the reputation of and respect for the program have increased in the community. Figure 11.2 demonstrates the rate of uptake over the ten years of the program.

Measurement of the Impacts of a Precautionary Program

A major problem for the EFP Program has been measurement of the impacts of the program on the environment. Given the successes of the uptake of the program and the actual level of implementation, one would expect to be able to measure such impacts, but there is no evidence from monitoring of the environment to demonstrate that the condition of the environment in general has improved as a result of the EFP. Indeed, in concentrated watershed-based studies of implementation of best practices in farming, the results of environmental monitoring have been inconclusive at best.

A significant problem in measuring precautionary practices is that most of the actions are taken in advance of an insult to the environment. Thus, the efforts prevent damage rather than repair it. How does one measure things that do not occur? This is a problem that is not dealt with well in the literature. A logical response is, "If damage is prevented, then the rate of decline of environmental quality should decrease or at least remain constant." Given the episodic nature of events that generate environmental damage, monitoring of these trends requires intensive long-term data to provide conclusive proof of the impacts of the programs.

Figure 11.1

Uptake of the Environmental Farm Plan (EFP) across Ontario

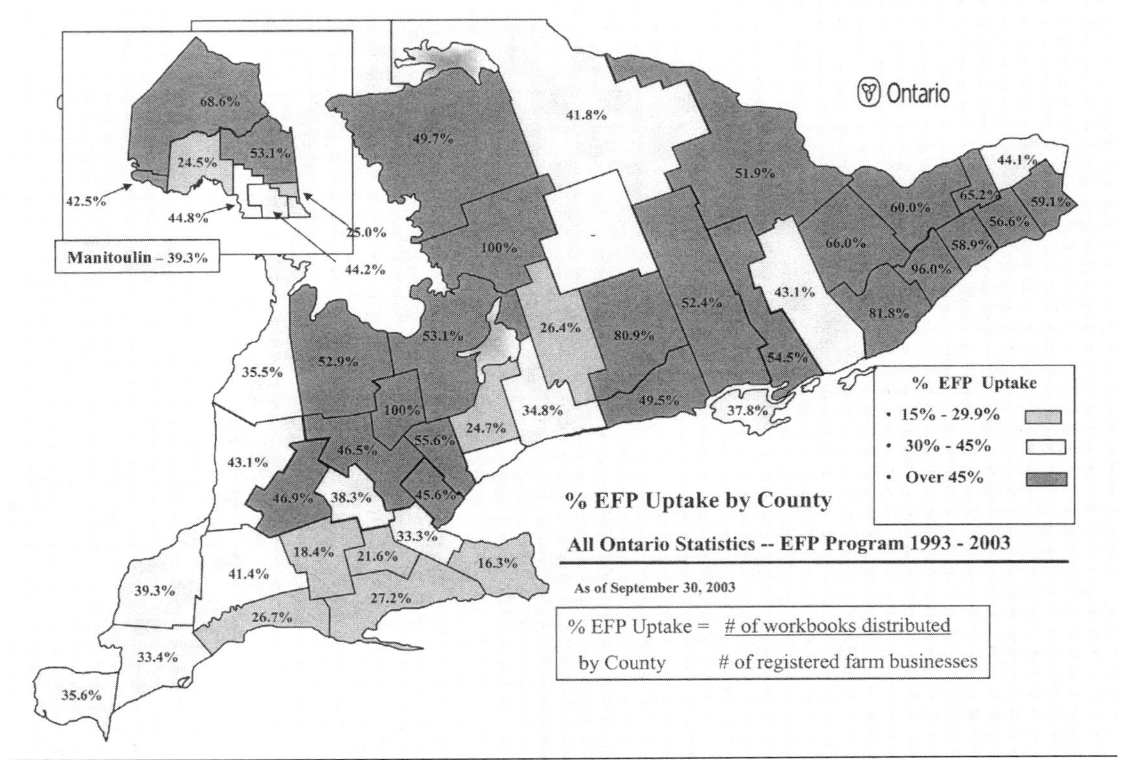

Sources: Ontario Soil and Crop Improvement Association, 31 July 2003; 2002 Ontario farm registration database.

Figure 11.2

Participation in the EFP in Ontario, 1993-2003

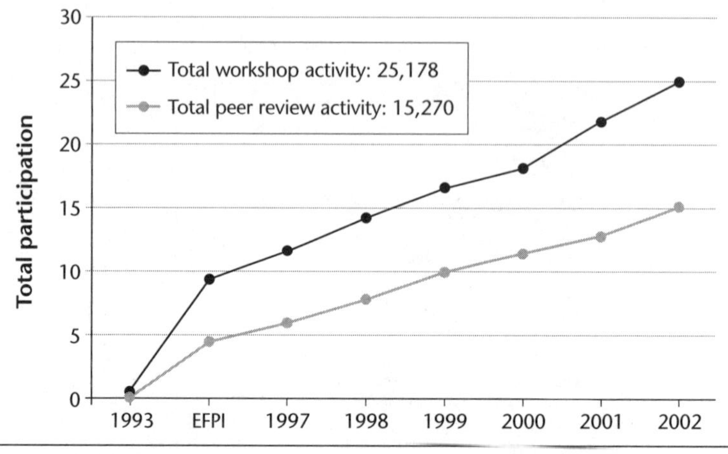

Source: Modified from H. Rudy, Ontario Soil and Crop Improvement Association.

In order to demonstrate the impacts of the EFP on the environment, we have had to resort to the use of a pathways model, which is based on an understanding of the process pathways by which an impact is delivered to the environment (causal linkages of impact delivery). In this study, the number of interventions that have reduced the opportunity for an impact to be delivered to the environment along a pathway has been used as the approach to identifying the impacts of risk aversion. An example is provided in Figure 11.3. Based on this analysis, half of the potential sources of risk (threats) for groundwater contamination identified in the farm plans have been acted upon to reduce the potential for contamination. Over time, this should result in a decrease in the rate at which ground water is degraded.

The problem with this approach is that ultimately there is a burden of proof that must be met for the public to continue supporting a program like the EFP. Thus, along with the program of risk aversion and mitigation, there must be an effective and convincing program for monitoring the quality of the environment. To date this has not been carried out.

The Changing Role of the EFP

At present the environmental farm planning process has been supported as a series of projects under the Agricultural Adaptation Council Program. While this has provided for the development and testing of the process, the process has gone beyond the status of a project and has indeed become a program that is an integral part of many other projects. The EFP has become a point of entry into many environmental protection programs that focus on

Figure 11.3

Pathway analysis of risk mitigation by sample farm plans

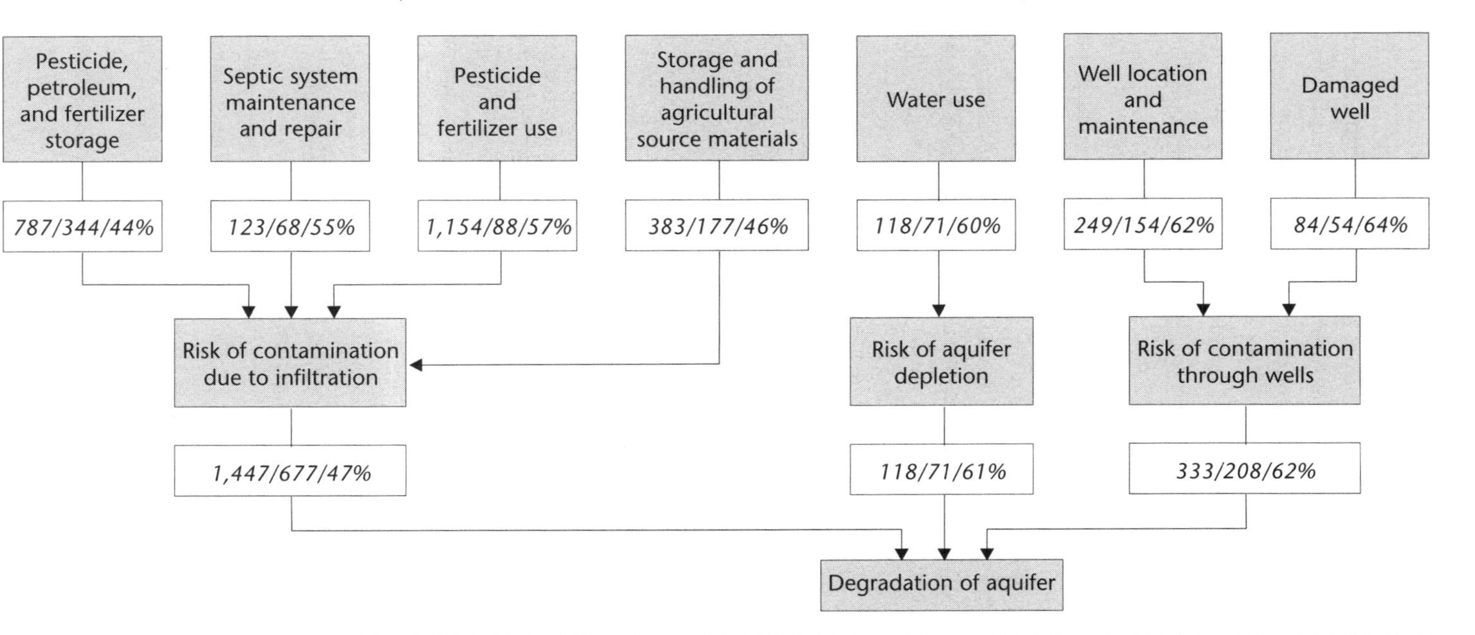

Source: Modified from FitzGibbon et al. 2000a.

water quality. It has also become a key component in supporting other measures such as nutrient management planning, planning for protection of species at risk, and planning for soil conservation and enhancement. All of these more focused tools for environmental management, some of which have regulatory requirements, use the EFP as a screening tool to direct the farmer towards more detailed management processes.

The EFP is not expected to take on the form of these management tools; rather, it should support the broader whole-farm system approach to agricultural stewardship of the environment.

The Environmental Farm Plan is evolving more along the lines of a broad risk audit such as that found in the ISO 14000 audit procedure. Much discussion of this has occurred, and preliminary attempts to relate the EFP to ISO 14000 have shown that the EFP is simpler, more logical, and more amenable than ISO 14000 to the diversity and complexity of farming systems. There are concerns that the cost required for ISO 14000 certification cannot be justified by the market advantage that might be gained (Wall et al. 1998). The EFP also has a number of features that have not been achieved by ISO 14000, including greater flexibility and adaptability, lower cost, and greater educational and community development value. On the other hand, ISO 14000 has features not available from the EFP as it is currently programmed, including on-site third-party review and continuous quality improvement (Wall et al. 1998). The linkage does, however, raise the issue of the use of the EFP as a tool for "branding" agricultural products as being produced in an environmentally responsible fashion. This has been used in a number of commodities to retain and enhance market share. A current pilot program using the EFP to promote Ontario beef is underway (2001-present). This change in the role of the EFP raises some issues for key aspects of the program. These include the focus on the educational value of the EFP, the confidentiality of the plan, and the shared responsibility of both the producer and the public for supporting environmental risk reduction. A further factor changing the role of the EFP is the ever-increasing presence of regulation as a means of requiring environmental management in agriculture.

The EFP and Regulation (the Relationship between Stewardship and Compulsion)

It is broadly considered that stewardship is an ethical action taken for both private and public benefit as a part of shared community values. Regulation, on the other hand, is viewed as the imposition of community values on those who disregard community norms and shared values to the detriment of the community and other individuals.

It is broadly agreed among environmental specialists that regulation is not a substitute for stewardship but should function to enhance and reinforce stewardship. The role of the EFP is one that lends itself to the defence of due care for those practices for which there is no regulatory standard.

There is an increasing trend towards regulation of agricultural practice. This began with initiatives in the European Union (Brower and Lowe 1998). The objective was to improve environmental conditions, especially water quality. Among the areas of regulatory focus are soil quality, water quality and quantity, air quality, biodiversity, and landscape. These protection measures were supported by funding programs under a series of regulatory measures contained in the Common Agricultural Policy for the European Union. In Canada the movement towards regulation is following some similar trends with nutrient management legislation, legislation to protect species at risk, and measures to protect rural quality in order to support tourism.

The advent of nutrient management legislation in Ontario and across Canada sets out one such code but deals with only one area in defining normal farm practice, compliance with which will protect farmers from nuisance and other challenges to their land use. There is a significant level of correspondence between farms that have used the EFP and those that have developed nutrient management plans (see Figure 11.4, based on Fitz-Gibbon and Thacker 2001). There is also a clear relationship between large livestock operations and regulation, as shown by a comparison between the distribution of large livestock farms and nutrient management bylaws in

Figure 11.4

Ontario farms completing Environmental Farm Plans (EFP) and nutrient management plans (NMP), by size of animal units, 2000

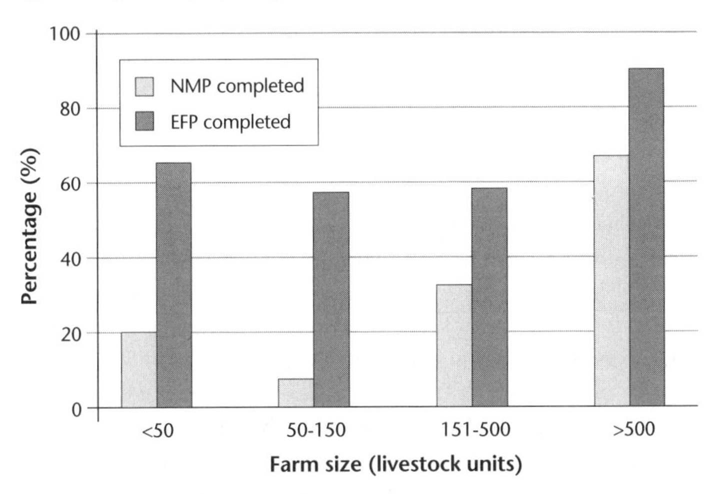

Source: Based on interviews taken at 150 sample farms in animal agriculture.

Figure 11.5

Locations of large livestock farms in Ontario and distribution of farms with completed EFP, 1996

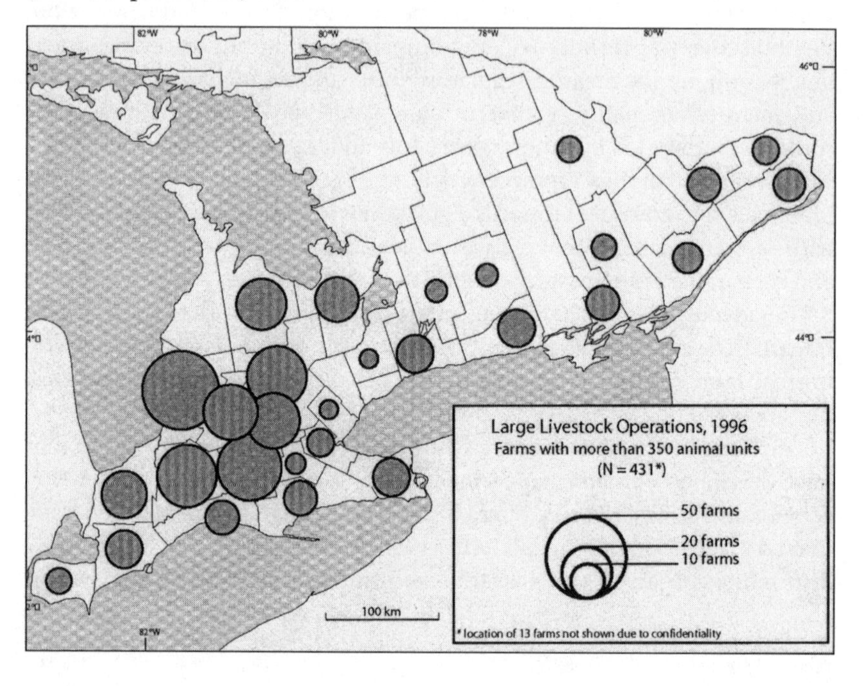

Ontario (see Figure 11.5). Thus, size has influenced the imposition of regulation, but environmental action is also clearly taking place alongside, if not in advance of, regulation.

The Environmental Farm Plan has gained general acceptance and provides some defence for activities not covered under various codes. The possible linkage of the EFP to regulation is clear since it has been an effective and efficient tool in identifying environmental perils that can be prevented, and it has achieved this success as a voluntary confidential process. The question that the linkage to regulation raises is how a voluntary, confidential process should respond to the compulsory publicly accountable process. This issue will be of concern where partnerships between private and public interests are the vehicle for resolving problems, a situation that is becoming more frequent as governments seek to utilize self-regulation as a means of complying with public policy. The report *Managing the Environment: A Review of Best Management Practices Executive Summary* (Executive Resource Group 2001) recommended that there be delegation of responsibility, but not necessarily accountability, for environmental management to nongovernmental organizations (NGOs) (strategic shift number 5). The report also recommends that the approach to environmental management incorporate integrated risk assessment and a commitment to continuous improvement in environmental performance. All of these measures are embraced by the EFP, but the problem of accountability remains. A movement towards greater accountability in the auditing of farm plans clearly will provide a more significant link between EFP and other regulatory codes such as the Nutrient Management Act.

The Future of the EFP Program

The success of the EFP Program has led to a number of major changes and innovations that will impact its future. The Agricultural Policy Framework, which is Agriculture and Agri-Food Canada's major initiative for the agricultural industry, will utilize the Environmental Farm Plan as the keystone component of its environmental programs. This means that a national program with a common concept and goal will be implemented in all parts of the country.

The EFP has provided a useful comprehensive environmental auditing tool for farming. It has had great success in raising awareness and as a tool for educating the agricultural community in recognizing environmental risk and implementing best management practices. However, the EFP now faces new challenges from both society and government. These challenges include accountability to the public for environmental performance, implementation of continuous quality improvement, integration with regulatory processes, and linkages to other management processes such as Hazard Analysis Critical Control Point (HACCP), Integrated Pest Management (IPM), and Nutritional Management Plans (NMPs). The transition in the role of the EFP from being an action taken by agriculture to protect the industry from expensive and overly intrusive regulation to being one of the components of a total quality management system for agriculture is a difficult one. This is especially so because of the desire on the part of agriculture for control of the process. If the changes that appear to be coming are to successfully integrate the EFP, it will be necessary to convince society that stewardship and precautionary management are a superior approach to regulation and reaction to environmental issues in agriculture.

Part 4: Lessons Learned

12
Integrating Farming Systems Analysis of Intensive Farming

Glen C. Filson and Chris Duke

Are these intensive farming systems sustainable in this era of globalization? Industrialization, globalization, and trade liberalization in a neoliberal era have led to the intensification of agriculture on the one hand and greater demands for the amelioration of its environmental and social effects on the other. Putting these issues in sharper focus, however, requires a careful review of some of our key concepts and findings. An interdisciplinary farming systems analysis approach has been employed to analyze these changing conditions. This provides us with a way of viewing the rural communities and their surrounding ecosystems.

Systems analysis also provides us with a way of synthesizing information and approaches from several disciplines in order to assess the environmental and socio-economic interactions and determine the most sustainable practices.

Environmental and Social Consequences of Intensive Agriculture

Agriculture and Agri-Food Canada (AAFC) has published national-level agricultural indicator analyses that raise environmental concerns about the effects of intensive agriculture (see Chapter 2). Like many watersheds in intensively farmed areas, the Grand River Watershed in southwestern Ontario is a mix of urban, rural, and natural areas. Manure nutrient loss is excessive in some fields. These losses tend to be diffuse, but tile drainage and proximity to watercourses can cause peaks in nutrient loads. The point sources of nutrient loads entering water bodies can be very difficult to characterize since they may be intermittent, one-time events, or coming from a very small location. Regardless of the source and cause, agricultural management can incorporate and respond to these known problems; sometimes it does, at other times it fails to do so. All the while, farmers' costs can vary from nominal to significant.

Some existing farming practices also contribute to reduced surface and groundwater quality, and so must be modified through the adoption of best management practices (BMPs). Across Ontario, well testing has shown that fecal coliform and nitrate levels affect many recently surveyed wells. Surface water soil sediment and nutrient loads vary by location and with time, but there are continuing water quality problems. Recently, bacterial content in the water has become a greater threat to human health than nitrates or pesticides. There are clearly different circumstances that are affecting point and nonpoint sources of water pollution that require improved management.

While soil management and soil cover, erosion, and carbon have improved, compaction has worsened. Lowering energy inputs, improving nutrient management, and sequestering organic carbon in soils is occurring and will increase farm efficiency and reduce greenhouse gas emissions over time while increasing soil fertility. Greenhouse gases continue to be produced by agricultural practices as well as by industry and transportation, so agriculture must adapt to the resulting climate change as well as play a role in their mitigation, such as through producing more renewable ethanol for fuel. Improvements in tillage implements have enabled Ontario farmers to reduce the risk of soil erosion in their fields. The increasing area of row crops, though, is partially offsetting these improvements. There are soils, such as clay soils, that are not conducive to current reduced tillage technology, but their soil erosion risks are not severe. There will undoubtedly be continued improvements in tillage systems that will be adopted if the agricultural commodity prices permit farmers to invest in long-term soil sustainability.

Biodiversity continues to decline because of the growth of intensive agriculture, nonfarm rural development, and encroaching urban sprawl. Farmers can improve the situation by reduced wetlands drainage and greater use of riparian zones, which also help to lower rates of disease and stabilize pest populations.

While farmers have increasingly adopted genetically modified organisms, many rural and urban people worry about their potential consequences over the long term. Unfortunately, we also lack adequate mechanisms for examining the use and impact of genetic modifications of plants or animals on farming systems.

To support a productive, sustainable, and economically viable agricultural production system, new approaches are clearly needed to have "clean" agroecosystems, and new policy tools are needed to support the drive to improve environmental management. To create these improved policy tools, we need the ability to analyze the socio-economic and biophysical linkages in order to develop new resolutions of the trade-offs between important economic, social, and environmental goals of farming systems. This will require a real dialogue and collaboration among farmers, government policymakers, and members of the physical and social sciences. But we need these tools now so that we can develop the capacity to face the increasingly rapid changes confronting agriculture and society.

The latest "food regime" has evolved simultaneously as corporate control of the food chains has become consolidated. This has facilitated further intensification in Canadian farm production, in turn causing the disappearance of many small family farm operations and exacerbating the hostility many small operators feel towards "usurping" commercial farmers. As exurbanites sprawl into the countryside, there has also been growing conflict between agriculturalists and nonfarm rural people. Many relative newcomers to rural Ontario want to adopt stricter regulations for farming systems, while many farmers continue to press for voluntary compliance, especially with environmental standards. Pressure to improve our food production systems and the health of our agroecosystems is rising steadily and ascendant.

Most consumers want high-quality food at low prices. The farmers' response, especially in North America, has been to adopt more intensive management of land, livestock, crops, agronomic inputs, and farm wastes. A rapidly growing but still relatively small group of farmers has opted instead for a range of alternative practices, from integrated to organic farming and various forms of community-shared agriculture.

Fewer but larger farm operations have become delinked from their rural communities. Increasing numbers of urbanites relocating into rural areas have often overwhelmed farmers politically at the municipal level. This conflict is epitomized by the desire of some municipalities to legislate minimum separation distances (MSDs) between newly constructed barns and the nearest neighbour. The Walkerton tragedy and the ensuing public inquiry has changed many things in Ontario, leading, for example, to the passage of the Nutrient Management Act legislating acceptable nutrient practices for all farmers except for the recent exemption given to relatively smaller operators.

In some parts of Ontario, farm-based rural communities have been in decline as the ranks of small-operation, independent farmers have thinned. Other rural communities have adapted in order to maintain their rurality by developing a variety of economic niches such as "boutiqueism" and agritourism to counter urbanization trends.

Aggressive expansion of urban industrialization into rural Ontario has been a major factor forcing changes within farming systems, but international competition in an era of freer trade has also pressured farmers to develop more intensive, consolidated, concentrated, and specialized farm practices. The hypothesis that the more industrialized our farms become, the worse rural quality of life will be has not yet been convincingly shown in either Ontario or Alberta as there are insufficient longitudinal studies. Agricultural decline in some rural communities has already been accompanied by healthy manufacturing or service endeavours that have absorbed former farmers in the process. There has been a steady inflow of urbanites into rural dwellings that have been severed from farms in southwestern Ontario. These exurbanites moving into rural areas tend not to participate in local community churches, athletics, and other forms of entertainment to the same degree as farmers, although the latter usually appreciate the contribution to the tax base that exurbanites make.

A shift in policy towards support for multifunctionality would enable farmers to be subsidized to maintain agricultural landscapes; funding would be provided for farmers' efforts to manage rural landscape sustainably such that biodiversity and other functions like agri-tourism can coexist with farming. Subsidizing the production of such noncommodity outputs as landscape and open space amenities, cultural heritage features, rural economic viability, and improved groundwater recharge could be a much more sustainable approach to rural land preservation than what we now have.

Farming systems analysis of the entire farming system and of the quality of life of the people part of this system helps transcend media sensationalism and broaden our understanding of the state of farming by adding social indicators to the environmental indicators that are predominant within universities and government. Surveys of farmers and nonfarmers show that environmental concerns are the major issues associated with the perceived quality of rural life. Whereas nonfarmers tend to favour a regulatory approach to dealing with these problems, such as that embodied in Ontario's Nutrient Management Act, farmers prefer a voluntary approach to introducing conservation BMPs. The economic benefits of more efficient agricultural production are low in priority for most members of the public, even though the economics of farming is the primary motivating factor for most farmers. This clash of interests explains much of the difference in perspective about the environmental consequences of intensive farming operations exhibited not only between farmers and rural nonfarmers but also between large, commercial operators and small farmers.

In social-class terms, there is growing differentiation among farmers due to the intensification of agriculture, further specialization, and the growing gap between relatively more corporate, commercial operators and the residual class of small, independent family farms. Farmers do not belong to one discrete social class, but are instead commercial farm owners, managerial farmers, farmers who are largely without employees, or part of the growing numbers of farm workers. Except for commercial farmers, most farmers need public assistance in order to achieve environmentally friendly production and credible nutrient management planning. Many of the more intensive, commercial farmers feel that they can afford to do the right things environmentally, need no help at all from government, and favour free trade at the same time. Whether they actually are environmental friendly depends on the farmers themselves.

A Framework for Farming Systems Analysis

Part 2 of this book develops a framework for conducting farming systems analysis of these environmental and socio-economic problems. These problems must be conceptualized and characterized within the context of sustainable agriculture. Consultation with farmers and other agricultural stakeholders plays an essential role in identifying problematic issues in the environment and rural communities. Indicators, measurements, costs, social concerns, and the capacity of the system need to be defined. Goals, objectives, problems, and possible solutions must also be identified collaboratively.

The establishment of physical and conceptual boundaries around the components is necessary prior to the specification of the indicators and other measures of the farming system. Boundaries should be imposed to make the analysis and discussion practical; at the same time, however, the indicators and the components should be chosen to permit the results of the analysis to be extended beyond the scope of the current analysis. Boundaries established around the problems help to define the context of the system being studied, and implicitly define parts of the system not currently seen to be relevant. This enables steps to be proposed to solve the problems identified. The progress and results need to be monitored using indicators and evaluation of other endpoints. Eventually, the outcomes of the process should be evaluated to aid in decision making and provide recommendations to improve the sustainability of the farming systems.

This can be facilitated by the use of a conceptual framework such as the Pressure-State-Response or Vigour-Organization-Resilience models of ecological system properties. Models that are formed of the system can be used to further characterize the system, evaluate problem-solving alternatives, and transfer this knowledge to other situations. The applications presented in Part 3 essentially follow these steps, but not in a strict recipe-like way.

Chapter 5 describes how predictive and summative indicators can be employed to understand the linkages between the socio-economic and environmental components of the farming system as well as between the farming system and the rural communities where they are found. How these predictive and summative indicators can work was illustrated generally with dairy farming systems in the chapter.

Indicators should be focused on the environmental and social goals and objectives of the project. The problem-solving framework, properties, and dimensions of sustainability are elaborated to show how farming systems researchers can develop integrated farm-level models that link changes in environmental quality with agricultural practices so that economic and environmental trade-offs can be quantified for policy analysis purposes. Some of the goals will be general, others spatio-temporally specific.

Integrating the needs of society and farmers, land capability, and environmental protection into a predictive framework is the challenge embodied in the assessment of these linkages. The authors of Chapter 5 argue for a system of predictive sustainability indicators at the farm level and summative sustainability indicators at the community level. There must be a statistical relationship between a community's predictive farm-level indicators and its cumulative summative indicators, although nonfarm activities also play an important role in the latter. Time-dependence of assessments of these indicators is an important factor that needs to be incorporated in models of the linkages between the farm and community's economic, social, and biophysical information flows, which in turn are connected by feedback loops.

Looking at the results of this work, summative indicators describe the state of the system and predictive indicators represent potential influences of the farming system on the community and environment. Some potential indicators could be found through open-ended questionnaires with the affected public. Linking indicator information to resource levels and locations, to community groups, and to farms and fields provides the critical linkage between subpopulations in the farming system. This points to who and what are affected by particular actions. Given the complexity of farming systems, the most important components and relationships have to be prioritized in order to understand the system.

Figure 5.2 provides a model of how the farming system components can be linked to the surrounding rural community as well as how the summative and predictive indicators are linked. While the inputs and outputs from the cropping and animal systems are fairly well developed, what the outputs do to particular summative indicators is not yet as well developed, in part because we do not yet have an adequate human and social model of decision making on farms. Whether this will ever be possible in a satisfactory way, given the complexity of the decision-making contexts, remains to be seen.

Chapter 6 takes the reader through the steps economics researchers must undertake when constructing a farm-level model that will account for environmental changes that occur with different agricultural practices so that the trade-offs for policy analysis can be quantified. The sustainability issues examined and accounted for in this modelling process will be of interest to researchers, farmers, and extensionists seeking to find ways of minimizing environmental damage and abatement costs. The authors reject the textbook solutions to this problem due to their lack of feasibility, in favour of the use of trade-off frontiers. The trade-off curves summarize environmental and economic health relationships for alternative farming systems so that BMPs can be identified.

Possible solutions to improve sustainability will involve trade-offs. These trade-offs will focus on discussion and goal setting when the farming system model is formulated. This helps groups identify tangible measures of sustainability affected by agriculture. These measures or indicators should be predictably affected by management decisions in order to help choose between trade-offs. Modelling the system and creating choice with trade-offs can require enormous amounts of data that likely change in space and time, but optimizing the model to achieve a desired outcome helps limit the range of possible choices of actions. In the case where several outcomes are sought, their relative importance can be prioritized. Modelling approaches that will solve for a number of optimal outcomes will help analyze likely outcomes given the priority of each indicator. If the outcomes are not deemed to be ideal, it would be necessary to re-evaluate the priorities of certain indicators. Given the range of different farming systems, it is difficult to construct generic models that will meet the needs of every problem solver. The collection of information on farming systems that will be applicable to specific fields, farms, and farm communities will remain necessary.

Applications of Farming Systems Analysis

Part 3 of this book presents farming systems research (FSR) applications of these and related methods. These farming systems applications provide essential solutions to many of the problems farmers face when trying to be productive and viable in the face of increasingly global competition, all the while doing this in such a way that the environment is protected. Following a systems approach to achieve several objectives for manure management, we have seen how modelling multiple possible systems can lead to the selection of the optimal solution for maximizing profit given the manure-handling system the farmer has.

Chapter 7 considers swine-finishing operations and presents advice to farm decision makers about how best to maximize the plant nutrient value of manure while avoiding the environmental damage that can result from mismanagement. The empirical model provides numerous options for different sizes of hog operations and for the available amount of cropland. This approach can be very useful for evaluating nutrient management plans and the alternatives a farmer may choose. When various scenarios are simulated, certain alternatives appear frequently, suggesting that these alternatives are practical under a range of farming systems. One highly valuable output of the model is the farmers' costs for each alternative along with the trade-offs for each. This provides information to policy makers who are deciding on support mechanisms for the adoption of improved environmental management and sustainability. Environmental care has a cost to farmers, and not all environmental objectives can be maximized at the same time.

The decision support system MCLONE4 shows how the processes of manure generation and management can be modelled with their costs. Chapter 8 builds upon the theme of balancing the environmental and economic concerns of manure management for both swine and dairy cattle by outlining how an on-farm computerized decision support program can be employed to environmental and economic advantage. The consequences to soil, water, crops, profit, and atmospheric quality are all modelled, and they are ranked from good to bad. The farmer has a tool with which to evaluate various approaches to manure management and can decide which alternative would best suit his or her priorities. The importance of manure management to the properties of sustainability is well known. This is why the MCLONE4 expert decision-making system and the Ontario Ministry of Agriculture and Food version, NMAN99 (Nutrient Management), with which it has been recently integrated, are so important.

Optimizing objectives or simulating various farming systems provides quantitative information on possible alternatives, trade-offs, and costs involved. Policymakers, environmental practitioners, and farmers have to make choices. This leads to opportunities to develop policies that support implementation of sustainable agricultural practices.

Chapter 9 discusses how when supply management was introduced three decades ago, there were some disruptive impacts, especially on small cheese-making rural communities. Now, however, unlike the *disappearing middle* of bimodally distributed very large and very small farms that characterizes California, Texas, and some other jurisdictions without supply management, it appears that the relatively good economic performance of those with as few as thirty to forty-nine milking cows shows that farm size is not the main factor creating economic success in Canada. The supply management system ensures that dairy farmers collectively govern their output to fit total milk demand. To prevent total milk output from expanding too fast under favourable pricing conditions, the Ontario Milk Marketing Board (now the Dairy Farmers of Ontario) was given the power to enforce milk production quotas on individual farms. The result has been stable farm gate milk prices, a more efficient milk transport system, the elimination of surpluses, and greater public support for the industry's need to maintain a steady supply of safe milk and milk products.

Additionally for dairying, there is a fairly flat average long-term cost curve that does not yield a substantially higher margin per extra commodity produced, as is often the case with the production of widgets in an industrial factory. Essentially, there is nothing preventing farm operations from getting bigger, especially if machinery and labour are available for the land to be worked. Despite this, many farmers and economists alike believe that the returns to scale entirely explain the precipitate growth of at least pork and poultry production. It may, however, have as much to do with the spread of a "treadmill" culture and a belief that one must "get big or get out."

Although also present within dairying, this is less common with Canadian milk production than with poultry and pork production, both of which have become much more concentrated and specialized. This may also signal a difference in the cultural perspective of the large operating poultry and pork producers, who, by putting so many more pigs and birds into barns than before, may often relate to their animals more mechanically than the relatively more medium or smaller sized dairy farmers do with their animals. In turn, animal welfare issues may be less problematic in dairy operations than in other, more concentrated livestock production.

Regardless of continuing threats from American and New Zealand dairy producers before the World Trade Organization, the continuation of this orderly marketing system appears to be both (1) assured in the short run and (2) a positive factor contributing to the overall sustainability of rural communities, especially in the Grand River Watershed of southwestern Ontario, where the study described in Chapter 9 was conducted. The authors conclude that while the *orderly managed* dairy farming system is not the most productive dairy farming system, it is a very viable system that provides room for a wide range of managerial styles.

Chapter 10 describes how a proactive set of farmers has been working to achieve a healthier watershed in Alberta's Crowfoot Creek Watershed. In this case, despite the fact that their work has generated many benefits to the stakeholders in adjoining subwatersheds, a substantial part of the costs of introducing BMPs is borne by the participating farmers. The latter have a relatively high level of awareness of the benefits of avoiding water contamination, in part because of the educational work of the program coordinator. The difficulty of identifying point sources of pollution in a watershed assessed as having poor water quality helps farmers avoid finger-pointing so that they can take collective responsibility for cleaning up the water. As the farmers begin to see the benefits of employing BMPs, their awareness of the importance of these BMPs grows. Also of considerable help have been the communication skills of the Crowfoot Creek Watershed Group's coordinator. The decentralized approach used has been easy to sell to the farmers, but it may be somewhat less effective environmentally. The voluntary nature of participation in this water quality group has been appreciated by these farmers and may provide an important model for other groups of this kind. Chapter 10 then provides some useful ideas about how cost sharing between the public and the farm community can improve the water quality within a watershed.

The final application of farming systems analysis uses a whole-farm approach towards participatory environmental management. Its authors discuss the history of the development of Ontario's Environmental Farm Plan (EFP). Once again, awareness and education have been important in spreading the EFP. The latest figures show that close to 40% of Ontario's farmers have participated in an EFP.

Farms that have carried out the EFP in Ontario tend to be the farms that have also developed nutrient management plans. These farms are also usually the largest farms, especially if they are livestock producers. These confidential, voluntary plans, which preceded the regulatory Nutrient Management Act, have become accepted as excellent tools for identifying environmental problems. Chapter 11 raises issues that arise from the relationship between voluntary cost-share programming and regulatory environmental programming. The authors note that "a movement towards greater accountability in the auditing of farm plans [such as embodied in the Nutrient Management Act] clearly will provide a more significant link between EFP and other regulatory codes such as nutrient management plans." This chapter's authors, including the chair of the Ontario Farm Environmental Coalition, John FitzGibbon, favour stewardship and precautionary management as a way of encouraging farmer participation in environmental farm planning in preference to compulsory environmental regulation.

The authors point to AAFC's Agricultural Policy Framework, a national program that seeks to establish nutrient planning and environmental management of farms throughout Canada. They observe that both society and various government levels are putting pressure on farmers to adjust to the consequences of some of their farming practices, particularly in the form of accountability. Within this context, the EFP *may* help to convince the public "that stewardship and precautionary management are a superior approach to regulation and reaction to environmental issues in agriculture."

General Observations and Conclusions

FSR work has often been criticized for its inability to move its analysis beyond the farm gate. We have tried to look outside the gate in order to determine some explicit policy consequences that flow from this work. For instance, the analysis of voluntary cost-sharing programs has clear implications for conservation authorities and agricultural environmental policymakers at the municipal, provincial, and federal levels.

Farming methods are changing and intensifying; so are the farming population and community. This will cause unexpected consequences for the sustainability of agriculture. Social and physical scientists need to be sharing their observations of these changes and working together to be aware of the new consequences that could arise. New indicators will be required and the use of past indicators re-evaluated.

We have seen in the latter chapters how a systems approach within a framework helps quantify the causes and effects of possible actions on all aspects of environmental sustainability. Intensive agriculture creates some concerns and provides some benefits. The popular media are generally critical of intensive agriculture because they often focus on one or two concerns. These operations are part of a system and should be evaluated in this context.

Agriculture will continue to intensify in the foreseeable future. The globalizing neoliberal agenda of our economic, social, and political systems encourages it. Alternative integrated and organic farming systems are also growing in popularity, mainly because of the negative environmental and social impacts that have been associated with modern, increasingly intensive agriculture, whether small or large. While we certainly may want to limit the concentration and centralization processes that culminate in factory-style farming, we have also seen that the smallest farm operations are often the ones least capable of producing and implementing effective nutrient management plans.

Comprehensive nutrient management and agroecosystem management from the field to the farm to the regional scale are increasingly coming to pass throughout Canada. Farmers are the agents of our farming system and they must be enthusiastic about their management. While we continue to demand that they produce more for less, we must also be prepared to support them with innovative means of achieving sustainable agriculture.

What have we said about intensive agriculture in Ontario and how the linkages may be followed to sustainability? First, we defined intensive farming as the tendency of increasing numbers of farms to have become relatively large operations that rely on a combination of mechanized forms of production and a small but growing number of agricultural workers. They represent relatively larger investments in land, labour, and capital than has traditionally been the case in Canada. AAFC's Environment Bureau has concluded that the growing intensity of agriculture increases environmental risk. An example is the increase in residual nitrogen levels resulting from more intensive livestock and the increased cropland now used to grow corn, soybeans, and other crops requiring high nitrogen levels. This increases the risk of nitrate contamination of water.

It remains to be determined whether the increased bacterial levels contaminating water supplies such as Walkerton's can be attributed entirely to more intensive agriculture or are simply due to the cumulative effects of small, medium, and large farms over time. These environmental risks as well as the incidence of bacterial contamination and excess phosphorus in soil and water may continue to grow with intensive agriculture, unless the public joins with farmers to fund conservation BMPs to lower environmental risks.

Many problems arise from the production and use of manure in intensive farming. The use of pesticides and herbicides has increased productivity but, although declining somewhat, the cumulative effect on the environment has also been problematic as farming has intensified.

We will have to develop methods to predict and analyze the side effects and long-term effects of the foods and therapeutics that may result from current genetically modified organisms (GMOs) and the second-generation ones with multi-gene traits. Because these genetically modified products will continue to be produced by an even more intensive form of agriculture than we have today, interdisciplinary farming systems research could play an important role in the assessment of their effects.

As we continue our FSR work at the University of Guelph, we are focusing our attention primarily on environmental protection, economic viability, and social acceptability. Voluntary cost-sharing programs have set an important but often politically difficult precedent, involving targeting farmers most in need of help to introduce conservation BMPs that minimize and control off-farm pollution from their operations.

Most conservation practices are not profitable for farmers even though they may be in the public interest because of their effect in protecting the environment. The cost of requiring *all* farmers to introduce various conservation practices when most farmers are in compliance with environmental standards or their farm land is not at risk due to its distance from water, lack of slope, and so on suggests that only a minority need to adopt these practices. The sites at greatest environmental risk can be identified using remote sensing and other assessments of environmental damage. In cases where it can be shown that societal benefits in terms of quality of life and environmental protection outweigh societal costs, the public subsidies now available in some jurisdictions ought to be expanded to encourage landowners to build improved manure storage facilities, develop buffer strips, fence cattle off from streams, and use variable input technologies to avoid excess application of nutrients.

We have learned that the properties of sustainability from productivity, viability, environmental protection, and social acceptability have specific dimensions for each farming system that can and must be monitored. The consequences for public health when environmental protection is not sufficiently guaranteed has been woefully demonstrated by the Walkerton tragedy.

Future Directions for Farming Systems Research in Ontario Agriculture

On-farm, pilot, and experimental research must be increased and the knowledge gained from such research must be transferred more effectively to other farmers. The intensification of certain segments of agriculture has been accompanied by the development of a sophisticated network of support agencies, service providers, and marketing systems. This is all taking place in a dynamic and increasingly populated rural area. These people and groups have a stake in the quality of the agricultural system and the quality of life in these areas. Farmers themselves have a huge stake and will be significant agents of change as they react to the economic, social, and environmental constraints. The systems approach to analyzing intensive agriculture enables informed decisions with a knowledge of the trade-offs that these decisions entail. All forms of agriculture are important to the economy and the country, and a holistic view of agriculture will assist in the creation of policies for a sustainable agricultural system for the benefit of our communities. The framework proposed in the book should permit greater consideration of the quality of life for farmers and nonfarmers alike in the management options proposed. Too much agricultural research has been limited to raising productivity and farm profitability without considering the impacts on the other sustainability properties of environmental protection and the social acceptability of farming practices. The debate between voluntary and imposed regulation is a manifestation of conflicts between the farm and the wider public over how the effects of increasingly intensive agriculture should be managed.

Regulatory policies being imposed by provincial governments may not greatly aid the process of providing direction or assistance for sustainability; rather, they may simply limit the ill effects of agriculture, intensive and traditional. Supportive, sustainable agricultural policies will have to be specific to the farming system and based on holistic, multiple methods and coordinated disciplinary and interdisciplinary approaches such as those outlined in this book.

This book shows how indicators can be used to provide measures of agricultural changes as well as how several modelling approaches (biophysical, economic, process-based, empirical, optimization, linear programming, and so on) can be combined to solve problems indicative of the lack of agricultural sustainability. Future work will require effort in several areas of FSR.

First, as we have seen, the choice of indicators will be re-evaluated, refined, and improved.

Second, modelling will gain much more importance. The indicators will have to be linked to parameters and variables in the farming system models. The models will be descriptive and predictive. Models for decision support from farm to regional level will help in formulating agricultural policy to a greater extent. For example, Agriculture and Agri-Food Canada is actively developing land quality modelling tools to assess land resource quality, environmental sustainability, and the economic impacts of current and proposed agricultural practices, policies, and programs.

Third, agroecosystem monitoring will be integrated with models, indicators, and near–real time data collection systems. Remotely sensed imagery data and geographic information systems will be needed to quantify agroecosystem variables in order to create "live" databases for use with the models. Indeed, we shall likely be able to conduct coordinated disciplinary research, sharing the same data sets collected systematically across the province in an integrated fashion.

In addition to nutrient management planning, it is time to look closely at establishing a broader form of relocalization of policy formation that extends public funds to farmers willing to undertake Environmental Farm

Plans. Should we not be paying our farmers to be good stewards of the land, as is done in many European countries? If people want safe, nutritious food and aesthetically pleasing rural landscapes with natural amenities, they must also be willing to support farmers' introduction of conservation BMPs.

Glossary

Agriculture and Agri-Food Canada (AAFC): "Agriculture and Agri-Food Canada provides information, research and technology, and policies and programs to achieve security of the food system, health of the environment and innovation for growth."[1]

Agriculture Non-Point Source Pollution (AGNPS) model: A management practice model geared towards assessing nutrient flow. For example, the AGNPS model can assess the effects of management choices such as fertilizer application rates on nitrate levels.

Agricultural intensification: Rising levels of capitalization through purchased nonfarm inputs combined with rising outputs per hectare.

Atrazine: A triazine herbicide used to control broadleaf and grassy weeds that can be found in corn, pineapple, Christmas trees, and other crops.[2]

Best management practices: Tools and techniques that are designed to restore and protect the environment. They may include, for example, the use of riparian buffer strips between crops and streams, or variable input technologies that ensure that excess nutrients are not applied to crops.

Blue baby syndrome: A blood disorder that occurs when a high concentration of nitrate is found in well water that is then used to prepare formula and other baby foods (Skipton and Hay 1998).

Capital cost: The purchase price for structures and equipment, such as manure handling equipment.

CENTURY model: A management practice model geared towards assessing nutrient flow. For example, the CENTURY model can examine the means to reduce the nitrates leaching into ground water from agriculture.

Complex systems: Systems that "are characterized by strong (usually non-linear) interactions between the parts, complex feedback loops that make it difficult to distinguish cause from effect, and significant time and space lags, discontinuities, thresholds, and limits. These characteristics all result in scientists' inability to simply add up or aggregate small-scale behavior to arrive at large-scale results" (Costanza et al. 1993, 545).

Crowfoot Creek Watershed Group (CCWG): Group made up of members, either agricultural producers or residents living within the Crowfoot Creek Watershed in Alberta, who take responsibility for water quality within their watershed.

Commercial farms: Often the most intensive farms, which not only employ farm managers and workers but also account for the majority of agricultural production. Commercial farms are more common in some commodities, such as tender fruit and vegetables, cut flowers, pork, and poultry.

---

1 Agriculture and Agri-Food Canada, "About agriculture and Agri-Food Canada," online at <http://www.agr.gc.ca/aafc_e.phtml> (retrieved 13 June 2003).

2 "Extoxnet ... pesticide information profiles," online at <http://extoxnet.orst.edu/pips/ghindex.html> (retrieved 13 June 2003).

Compliance-Diagnosis-Warning Typology (CDWT): An example of a conceptual framework for assessment that emphasizes system conformity to standards.

Concentration in farming: Occurs when fewer, larger farms develop with increasing linkage of these farms with food processing as these larger farms develop guaranteed markets for their produce and processors obtain assured supplies of produce (Bowler 1992).

Conservation tillage: A soil conservation method that enables crops to be grown with minimal cultivation of the soil (Peet 2001).

Coordinated disciplinary research: Research that involves researchers from relevant disciplines who interact primarily in the planning and analysis phases of the research but work independently during the actual research phase itself, using all the discipline-specific tools at their disposal.

Cost-benefit analysis: A quantitative technique used to determine the possibilities of a project, program, or plan by measuring the costs and benefits. Its purpose is to promote the most efficient allocation of resources through informed decision making. It depends on the criterion of whether or not a project can be justified based on benefits minus costs.

Dairy Forage Simulation Model (DAFOSYM): A simulation model that evaluates and compares the economic and environmental impacts of alternative productive systems for dairy farms (Rotz and Gupta 1995).

Decision support system (DSS): Software developed in order to provide managers with information that helps them make decisions. MCLONE4, a CD-ROM available from the Canadian Farm Business Management Council, is decision support software designed to help farmers conduct manure management so that their profitability is maximized while environmental risks associated with livestock production are minimized.

Density Equalized Map Projection (DEMP): An example of stochastic programming that focuses on the geographic distributions of something like a disease, which has the advantage of being able to represent within-year tactical decision making, and of explicitly representing probabilities of adverse environmental outcomes.

Dicamba: A benzoic acid herbicide that can be applied to leaves, soil, or certain crops (US National Library of Medicine 1995).

Disappearing middle phenomenon: A phenomenon characterized by a bimodal distribution of farms where there are some large farm operations, a large number of small farm operations, and few medium-sized farms.

Driving Force–Outcome–Response model: By "driving force" the authors of this model refer to the forces influencing agricultural activities. The outcomes are the environmental consequences of agriculture. "Response" refers mainly to societal responses to any changes that occur in the driving forces or the outcomes with respect to what technologies farmers adopt and how consumers and governments react.

Ecological space (environmental capacity): "A term used to describe what each person on earth would be entitled to if a sustainable amount of the planet's resources were distributed equally."[3]

Econometric model: An empirical model used to estimate the factors of economic systems. In the context of this book, econometric models are statistical representations of farm-level systems, often estimates as aggregate systems of equations for input demand and output supply derived from duality theory.

Environmental Farm Plan (EFP): The Farm Plan refers to documents that are voluntarily prepared by farm families to raise their awareness of the environment on their farm. "Through the EFP process, farmers will highlight environmental strengths on their farm, identify areas of environmental concern and set realistic goals and timetables to improve environmental conditions" (OMAF 2003d, 1).

Eutrophication: The increase in nutrient status of natural water that causes the accelerated growth of algae or water plants, depletion of dissolved oxygen, increased turbidity, and general degradation of water quality.

---

3 Japan Center for a Sustainable Environment and Society (JACSES), "Ecological SPACE (Sustainable Production and Consumption) Program," online at <http://www.jacses.org/en/ecosp/index.html> (retrieved 12 June 2003).

Evaluation: The specific meaning of this within farming systems research involves the steps that determine the success of the project by generating information and evaluating its impacts, improvements, and effectiveness.

Factory farming: When farms become entirely industrialized and owned by corporations; consist of hundreds or thousands of contained farm animals; routinely use antibiotics, hormones, and other chemicals to promote rapid animal growth; and place the quantity of food produced and profits obtained ahead of environmental concerns or sometimes human safety, the farms are often denigrated as factory farms.

Fecal coliform bacteria: A type of bacteria from manure that can be transported into ground water and cause watershed contamination.

Farming systems: "Farming systems involve a complex combination of inputs, managed by farming families but influenced by environmental, political, economic, institutional and social factors" (NRI-ORG, Farming Systems, 28 January 2004).

Farming systems research (FSR): A way of assessing specified elements of farming systems, making predictions about the effects of those systems based on the implications of change both within systems and beyond their boundaries, and evaluating these conditions and possible responses in the context of the goal of sustainability.

Food regime: The global food network that is tied into all the regions of the world. Three regimes have resulted from the intensification and industrialization of farming. The first food regime was during the European integration of colonies into domestic European economies. The second regime was created following the Second World War and was dominated by American agribusinesses. The latest food regime is a result of globalization, liberalization of trade, and a wave of privatization that has increased in the global trading of food (Frayne 2002).

Genetically modified organisms (GMOs): Gene manipulation that can help increase crop production (Harwood 1990). The process combines genes from organisms through gene technology. The resulting organisms are called genetically modified, genetically engineered, or transgenic. Thus, genetic modification is a special set of technologies that alter the genetic makeup of living organisms such as plants, bacteria, or animals.[4]

Globalization: "The accelerated integration of capital, production, and markets globally driven by the logic of corporate profitability" (Bello 2003).

Goal programming (GP) model: An applied modelling approach that allows for more than one objective to be optimized simultaneously through the specification of targets for each goal, with the mathematical objective being to minimize deviations from those targets.

Goldschmidt Hypothesis: Hypothesis stating that the more industrialized our farming becomes, the worse the rural community quality of life and socio-economic conditions will become.

Greenhouse gases: Emissions that include carbon dioxide, methane, and nitrous oxide, among others. These gases prevent the sun's ultraviolet rays from being reflected back into the atmosphere, causing the earth's surface to warm up.

Holistic agriculture: Includes low-input and sustainable agriculture (LISA), ecological alternative agriculture, and organic agriculture, all of which share an ecosystem-oriented instead of a world market–oriented vision.

Hydraulic loading limit: Maximum rate of liquid manure that can be applied without causing runoff.

Integrated Pest Control (IPC): A management process that uses different kinds of pest control tactics to help manage disease, insects, weeds, and animal pests.

Integrated Pest Management (IPM): A program designed to help manage diseases, insects, weeds, and animal pests through pest management recommendations that minimize the need to use chemical pesticides (OMAF 2003b).

---

4 Human Genome Program, US Department of Energy, "Genetically modified foods and organisms," online at <http://www.ornl.gov/TechResources/Human_Genome/elsi/gmfood.html> (retrieved 13 June 2003).

Intensive agriculture: Type of agriculture that employs relatively larger investments in land, labour, and capital such that larger farm operations continue to grow, displacing many smaller farms.

Intensive livestock operations (ILOs): Intensive cattle, hog, and poultry farming in large operations, where manure and waste disposal are controlled.

Interdisciplinary research: Research that involves a high degree of interaction in planning, conducting, and analyzing the results, among researchers who are disciplinary specialists in both the social and biophysical sciences.

ISO 14000: International Organization for Standardization environmental auditing system that certifies an organization's environmental management program.

Kyoto Protocol: An agreement among many countries to accept treaty rules and cut back on their emissions of carbon dioxide and other global warming gases to a level 6% below 1990 levels by the year 2010. The Kyoto Protocol has not yet been accepted by all countries, including the United States, although Canada is attempting to implement it domestically.

LEACHP (Leaching, Estimation and Chemistry, Pesticide) model: A management practice model that is suited to examining pesticide movement.

Learning platforms: Mechanisms for enabling decision making to occur within a bioregion. They may include environmental groups and conservation specialists who work with farmers who require remediation measures, and can be very effective in promoting environmental management practices.

Linear programming: A quantitative technique that enables models to be developed that find optimal solutions and select mixtures of systems components to maximize objective functions such as maximizing profits while protecting the environment.

Margin of error: The amount of space for you to make errors. The larger the margin of error, the more room you have to make mistakes.

Mathematical programming (MP) models: Optimization techniques that seek to solve real-world problems. MP models seek the best possible outcome to a broad range of biophysical, environmental, and economic circumstances.

Maximizers: Producers who are driven to maximize profits.

MCLONE (Manure, Cost, Labour, Odour, Nutrient Availability, Environmental Risk): A decision support software package that combines scientific insights from several disciplines into an expert system for manure management designed for farmers. It can evaluate solid or liquid dairy and swine manure systems and solid poultry manure systems, and includes weather, crop, soil, and economic data for different regions of Ontario.

Milk Act: An Ontario law that compels dairy farmers to collectively limit the industry's output of milk to what the market will bear at a price calculated by formula to cover the cost of production.

Minimization of Total Absolute Deviations (MOTAD) model: An empirical risk model that is easily able to incorporate empirical risk distributions or distribution parameters in order to evaluate the trade-off between technical and/or market risk and return (often measured in profit).

Minimum separation distances (MSDs): The smallest distance that certain agricultural activities, such as new hog barns, must be from buildings or public areas.

Mixed integer programming (MIP): A slightly more complicated extension of the linear programming models, where MIP sets selected activities (columns in the models) as integer variables.

Nearly optimal linear programming (NOLP): An applied modelling approach producing solutions that are not most favourable (optimal) with respect to any one objective but instead are somewhat or "nearly" most favourable (optimal) for all objectives.

Nutrient Management Act (NMA): An Ontario law that sets out standards for nutrient management on farms while protecting the environment (OMAF 2003c).

Nutrient management plan (NMP): The objective of NMPs is to use nutrients such as phosphorus, nitrogen, and potassium in a proper manner for the best possible economic benefit, while minimizing the impact on the environment (OMAF 2003a).

Odour rating/risk: One of MCLONE4's functions based on the estimated odour level at the nearest neighbour, while the system odour rating is based on the distance required to obtain proper odour dilution (a minimum separation distance).

Ontario Farm Environment Coalition (OFEC): A group consisting of most major agricultural organizations in Ontario. It has the formal membership and structure of a general forum as well as a steering committee and various working groups to deal with specific issues.

Ontario Ministry of Agriculture and Food (OMAF): The Ontario government ministry that provides programs and services to help address present and future needs for the province's agricultural and food industries. Its acronym used to be OMAFRA before 2003, when it was also in charge of Rural Affairs; OMAF has ceded control of Rural Affairs to the Ministry of Municipal Affairs and Housing.

Optimal residual level: The optimal residual level enables both environmental and economic health variables to be expressed in monetary terms. Conversion of the physical impacts of agricultural practices on the environment into monetary values permits a direct comparison of the off-farm environmental costs and the on-farm abatement costs.

Optimization model: A model that involves the specification of a behavioural assumption (e.g., profit maximization). Optimization and simulation models are both systems of equations and/or inequalities designed to replicate farm-level activities related to production, marketing, finance, etc.

Organic farming: Organic farming produces such commodities as grains, produce, dairy, and other products without the use of chemicals such as pesticides and fertilizers.

Pesticide: A form of insect repellent. The word "pesticide" can be used in a very general sense to include herbicides, fungicides, and insecticides.

Phytase: "An enzyme that breaks down the undigestible phytic acid (phytate) portion in grains and oil seeds; thereby, releasing digestible phosphorus and calcium for the pig."[5]

Precautionary principle (PP): A management approach where the actions taken are proactive, rather than reactive, responding to a disaster after it has occurred.

Predictive indicators: Indicators for which valid measurable data that can be obtained at the farm level, such as NH_3 losses from barns, farm-family income, and the surplus phosphorus load. Predictive indicators represent potential influences of the farm on the community. These influences are seen by their effect on the summative indicators. Linkages between the farm and community can be thought of as the cause-and-effect or statistical relationship between the predictive and summative indicators (see *Summative indicators*).

Pressure-State-Response model: An example of a conceptual framework for assessing the sustainability of an ecosystem; it emphasizes system processes.

Principle of proportionality: A management tool that stipulates that actions be taken in proportion to the risk of damages that might occur.

Recombinant Bovine Somatotropin (rbST): "Recombinant Bovine Somatotropin (rbST) is a veterinary drug [for the use of dairy cattle], produced through biotechnology, which is not approved for use in Canada and has been under review in Health Canada for the past nine years" (Health Canada, October 1998).

Response-Inducing Sustainability Evaluation (RISE): A basic tool to assess the sustainability of farms, "based on twelve indicators for the economic, ecological, and social situation: energy consumption, water consumption, situation of the soil, biodiversity, emission potential, plant protection, wastes and residues, cash flow, farm income, investments, local economy, social situation of farmer family and employees. For each indicator the 'Driving force' (D) and the 'State' (S) are assessed" (Häni 2002).

Riparian buffer strips: Zones of ecological compensation (ZECs) along the margin of waterways, such as hedges with herbs and native flora, that can provide the resources needed for beneficial organisms to develop, which can help keep pests and disease in check.

Riparian management (RM): Management that provides habitat for birds, fish, and other animal life between farmed land and waterways.

---

5 L. McMullen and P. Holden, "Phytase fact sheet," US Department of Agriculture (USDA), online at <http://extension.agron.iastate.edu/immag/pubs/phytase.doc> (retrieved 15 June 2003).

Rural Water Quality Program (RWQP): A set of financial and technical incentives for landowners to conduct best management practices designed to improve agroecosystem health while reducing water treatment costs developed by the Grand River Conservation Authority for the Grand River Watershed.

Satisficers: Producers who are satisfied with a reasonable income.

Scoping: A defining feature of most environmental impact assessment frameworks, where valued environmental components and associated issues are defined and carried forward into the assessment process.

Simple commodity production (SCP): Production by farmers and other small non-farm operators whose main purpose is to produce for their own subsistence, with the rare use, if any, of hired employees. Most of their production does not become a commodity, unlike that of most contemporary Ontario small farm operators. True simple commodity producers sell or exchange the excess they produce beyond what they consume themselves.

Simulation: Bellinger (2004) defines simulation as "the manipulation of a model in such a way that it operates on time and/or space to compress it, thus enabling one to perceive the interactions that would otherwise not be apparent because of their separation in time or space. This compression also provides a perspective on what happens within the system, which, because of the complexity of the system, would probably otherwise not be evident."

Specialization in farming: Specialization occurs as more of the total farm or regional output is accounted for by particular products (Bowler 1992).

Stakeholders: All of the individuals who are affected by the planning and implementation of a program.

Stewardship: An ethical action that is taken for both private and public benefit as a part of shared community values.

Subsistence farming: Family farming on relatively small parcels of land or the least fertile soil. Most of the food produced is for the family's own survival.

Summative evaluation: An evaluation process that normally occurs at the end of a program and that focuses on the impact of the program. For example, as an assessment of the sustainability of the farming system, summative evaluation is used to determine the extent to which the farming system is profitable, protects the environment, is socially acceptable, and is reasonably equitable for men and women and all social classes affected by the farming.

Summative indicators: Indicators that the community applies to entities, substances, or factors such as water quality, bacterial concentration, and value of milk quota. Predictive indicators at the farm level have a potentially cumulative impact on summative indicators at the community level (see *Predictive indicators*).

Supply management system: A system that controls the total supply of product from a farming system as a way of controlling the price of the product. Within Canada's dairy system, it is a process that enables dairy farmers to collectively govern their output within the confines of total milk demand so that they receive a reasonable profit for the milk that they produce.

Sustainable agriculture: Term used to refer to production systems that are environmentally benign (or enhancing), economically viable, and socially acceptable. Five system-level attributes form the basic elements of sustainability: productivity, security, protection, viability, and acceptability.

Synthetic lysine: A dietary treatment consisting of one of the essential amino acids in the pig's diet. The added lysine can replace some of the crude protein in the ration, thereby reducing nitrogen and phosphorus emissions in the hog manure.

Systems analysis: A mechanism for understanding and integrating biophysical and socio-economic relationships among different spheres of a system. Systems focus on limited aspects of reality with clear boundaries and an arrangement of component parts that continuously interact to achieve goals by transforming inputs into outputs.

Systems approach: Concentrates on the relationships between different parts of a system or entity which maintains itself through the interaction of its part. Systems function in time and space (Bellinger 2004).

System Dynamics method: A method developed by Forrester (1968) for industrial systems. It assumes that a system has a goal, and its behaviour over time to reach that goal is controlled by feedback loops. The feedback loops are an information flow that represents the input for controlling the rate of change of the various model component outputs. The approach therefore relies mainly on the level or stock and on the rate or flow.

Trade-off curves: A convenient means of summarizing information on the relationship between environmental and economic health for alternative agricultural production systems and issues regarding sustainability, so that best management practices can be identified. "The results are presented in the form of trade-off curves that are intuitive and easy-to-understand for policy makers using the econometric production models estimated on observed behaviour of the population of farmer[s]" (Environmental Sciences Group, 29 January 2004).

Trade-offs: A balancing between two or more phenomena, such as the profitability of an enterprise and the negative environmental consequences of that enterprise. The analysis of trade-offs within agriculture attempts "to quantify trade-offs between key sustainability indicators under alternative policy and technology scenarios" (Environmental Sciences Group, 29 January 2004).

Vigour-Organization-Resilience: A conceptual framework for assessment that invokes certain ecological properties of systems.

Whole-farm modelling: David Pannell (a co-author of Chapter 6) describes whole-farm modelling as follows: "Simulation approaches to whole-farm modelling range from very simple to extremely complex. Simple simulation models are common and widely used. Most farm advisors and many farmers build simple whole-farm budgets, which are in essence simple simulation models, with almost all of the detail simplified away. They can be very valuable and revealing, especially in the hands of an experienced farmer or advisor (Malcolm). At the other end of the complexity spectrum, is an integrated system of bio-physical dynamic simulation models (one for each species of plant or animal on the farm), perhaps feeding directly into an economic model" (online at <http://www.general.uwa.edu.au/u/dpannell/dpap961f.htm>).

Zones of ecological compensation (ZECs): In order to support sustainable practices, corridors or ZECs of hedgerows, riparian strips and other near-natural zones can be established for the purpose of ecological compensation (Swiss Agency for Environment, Forests and Landscape, 5 June 2003).

References

Books, Articles, and Reports

Abdallah, C.W., and T.W. Kelsey. 1996. Breaking the impasse: Helping communities cope with change at the rural-urban interface. *Journal of Soil and Water Conservation* (November/December): 462-66.

Acton, D., and L. Gregorich, eds. 1995. *The health of our soils: Toward sustainable agriculture in Canada*. Ottawa: Centre for Land and Biological Resources Research, Agriculture and Agri-Food Canada.

Addiscott, T.M., and R.J. Wagenet. 1985. Concepts of solute leaching in soils: A review of modeling approaches. *Journal of Soil Science* 36 (3): 411-24.

Agriculture Canada. 1993. Developing environmental indicators for agriculture. Environmental Indicator Working Group Discussion Paper. Ottawa: Agriculture Canada.

Allen, T., and T. Starr. 1982. *Hierarchy perspectives for ecological complexity*. Chicago: University of Chicago Press.

Altieri, M.A., K. Letourneau, and J. Davis. 1983. Developing sustainable agroecosystems. *BioScience* 33 (1): 45-49

American Society of Agricultural Engineers. 1998. Manure production and characteristics. In *ASAE Standards*, 45th ed., 646-48. St. Joseph, MI: ASAE.

Antle, J.M. 1987. Econometric estimation of producers' risk attitudes. *American Journal of Agricultural Economics* 69 (3): 509-22.

–. 1988. *Pesticide policy, production risk and producer welfare: An econometric approach to applied welfare analysis*. Washington, DC: Resources for the Future.

Antle, J.M., and R.E. Just. 1991. Conceptual and empirical foundations for agricultural environmental policy analysis. In *Commodity and resource policies in agricultural systems*, ed. R.E. Just and N. Bockstael, 97-128. New York: Springer-Verlag Publishing.

Antle, J.M., S.M. Capalbo, and C.C. Crissman. 1998. Tradeoffs in policy analysis: Conceptual foundations and disciplinary integration. In *Economic, environmental and health tradeoffs in agriculture: Pesticides and the sustainability of Andean potato production*, ed. C.C. Crissman, J.M. Antle, and S.M. Capalbo, 21-40. Boston: Kluwer Academic Publishers.

Arcand, C., A. Corregian, L. Hunter, L. King, C. Thomas, and J. Van Campen. 1999. Agrifood problem solving: Implications of large animal units. Final report, University of Guelph, Department of Agricultural Economics and Business.

Baker, D. 1993. Inability of farming system research to deal with agricultural policy. *Journal of Farming Systems Research-Extension* 4 (1): 67-86.

Baltussen, W.H., and A. Hoste. 1993. *Consequences of reducing ammonia emissions for pig farms*. Den Hag, Netherlands: Landbouw Economisch Instituut.

Bangay, G.E. 1976. *Livestock and poultry wastes in the Great Lakes Basin: Environmental concerns and management issues*. Social Science Series 15. Burlington, ON: Environment Canada.

Barnett, G.M. 1994. Manure P fractionation. *Bioresource Technology* 49: 149-55.

Barry, D. 1998. *Linkages in simulation modelling of farming systems: Literature review.* Guelph: University of Guelph Farming Systems Group.

Bawden, R. 1995. On the systems dimension in FSR. *Journal of Farming Systems Research-Extension* 5 (2): 1-18.

Bellinger, G. Simulation. Online at <http://www.systems-thinking.org/simulation/simulation.htm> (retrieved 7 April 2004).

Bello, W. 2003. The stalemate in the WTO and the crisis of the globalist project: Update on the World Trade Organization and global trends. Paper presented at the Hemispheric and Global Assembly against the FTAA and the WTO, Mexico City.

Benbrook, C. 1991. Introduction. *Sustainable agriculture research and education in the field,* by National Research Council. Washington, DC: National Academy Press.

Benbrook, C., and F. Mallinckrodt. 1994. Indicators of sustainability in the food and fiber sector. Paper prepared for the Sustainable Agriculture and Rural Development Forum (SARD).

Berdequé, J. 1993. Challenges of farming systems research and extension. *Journal of Farming Systems Research-Extension* 4 (1): 1-10.

Berkes, F., S. Cairns, G. McBean, G. Peterson, F. Westley, and N. Yan. 2003. Where uncertainty is high and controllability low: A research agenda for new frontiers in ecosystem management. Paper presented at CIAR Workshop, Montreal.

Beus, C., and R. Dunlap. 1990. Conventional versus alternative agriculture: The paradigmatic roots of the debate. *Rural Sociology* 55 (4): 590-616.

Biggs, E., ed. 1990. *History of milk marketing in Ontario: Research materials, 1930-1990.* Mississauga: Ontario Milk Marketing Board.

Binswanger, H.P. 1980. Attitudes toward risk: Experimental measurement in rural India. *American Journal of Agricultural Economics* 62 (1): 395-407.

Bird, E., T. Edens, F. Drummond, and E. Groden. 1984. Design of pest management systems for sustainable agriculture. In *Sustainable agriculture in temperate zones,* ed. C.A. Francis et al., 55-110. New York: Wiley.

Boland, M.A., P.V. Preckel, and K.A. Foster. 1999. Economic analysis of phosphorus-reducing technologies in pork production. *Journal of Agricultural and Resource Economics* 23 (2): 468-82.

Bond, G., and B. Wonder. 1980. Risk attitudes amongst Australian farmers. *Australian Journal of Agricultural Economics* 24 (1): 16-34.

Bowler, I.R. 1992. *The geography of agriculture in developed market economies.* New York: Wiley.

Boyd, D. 2003. *Unnatural law: Rethinking Canadian environmental law and policy.* Vancouver: UBC Press.

Boyle, K.J., G.L. Poe, and J.C. Bergstrom. 1994. What do we know about ground water values? Preliminary implications from a meta-analysis of contingent valuation studies. *American Journal of Agricultural Economics* 76 (5): 1055-61.

Bradshaw, B., and B. Smit. 1997. Subsidy removal and agroecosystem health. *Agriculture, Ecosystems and Environment* 64: 245-60.

Brower, F., and P. Lowe, eds. 1998. *CAP and the rural environment in transition: A panorama of national perspectives.* Wageningen, Netherlands: Wageningen Press.

Bryden, J. 1994a. Interactions between the farm household and the community: Effects of non-agricultural elements in farm household decision-making on farming systems. In *Rural and farming systems and analysis: European Perspectives,* ed. J.B. Dent and M.J. McGregor. Wallingford, UK: CAB International.

–, ed. 1994b. *Towards sustainable rural communities.* Guelph: University of Guelph.

–. 2002. Multifunctionality, agriculture and rural development: A European perspective. Talk given at the University of Guelph, Guelph, ON, 23 March.

Bucknell, D., G.C. Filson, and S. Hilts. 2003. Farm and non-farm attitudes to environmental practices in agriculture in two subwatersheds of Ontario's Grand River. Unpublished manuscript, University of Guelph.

Buttel, F.H. 1983. Farm structure and rural development. In *Farms in transition,* ed. D.E. Brewster, W.D. Rasmussen, and G. Youngberg. Ames: Iowa State University Press.

Buttel, F.H., G.W. Gillespie Jr., O.W. Larson III, and C.K. Harris. 1981. The social bases of agrarian environmentalism: A comparative analysis of New York and Michigan farm operators. *Rural Sociology* 46 (3): 391-410.

Cairns, J., P. McCormick, and B. Niederlehner. 1992. A proposed framework for developing indicators of ecosystem health. *Hydrobiologia* 263: 1-44.

Caldwell, J., and A. Christian. 1996. Reductionism, systems approaches, and farmer participation: Conflicts and contributions in the North American Land Grant System. *Journal for Farming Systems Research-Extension* 6 (2): 33-44.

Caldwell, W. 1994. Consideration of the environment: An approach for rural planning and development. *Journal of Soil and Water Conservation* 49 (4): 324-32.

–. 2000. Livestock and agricultural intensification: Community perceptions of environmental, economic and social impacts as an impediment to agricultural production. Proposal approved by the Ontario Ministry of Agriculture, Food and Rural Affairs.

Carson, R. 1962. *Silent spring.* Thorndike, ME: G.K. Hall.

Castle, G. 1993. Agricultural waste management in Ontario, Wisconsin and British Columbia: A comparison of policy approaches. *Canadian Water Resource Journal* 18 (3): 217-27.

CCIAD (Climate Change Impacts and Adaptation Directorate). 2002. *Climate change impacts and adaptation: A Canadian perspective.* Ottawa: Natural Resources Canada.

Chambers, R. 1995. *Poverty and livelihoods: Whose reality counts?* Brighton, UK: Institute of Development Studies, University of Sussex.

Charnes, A., and W.W. Cooper. 1959. Chance constrained programming. *Management Science* 6:73-79.

Chopra, V.L., V.S. Malik, S.R. Malik, and S.R. Bhat. 1999. *Applied plant biotechnology.* Enfield, NH: Science Publishers.

Cochrane, W.W. 1958. *Farm prices, myth and reality.* St. Paul: University of Minnesota Press.

Conway, G.R. 1987. The properties of agroecosystems. *Agricultural Systems* 24: 95-117.

–. 1991. *Unwelcome harvest: agriculture and pollution.* London: Earthscan.

Cooke, G.W. 1982. *Fertilizing for maximum yield.* 3rd ed. New York: Macmillan.

Costanza, R., B. Norton, and B. Haskell, eds. 1992. *Ecosystem health: New goals for environmental management.* Washington, DC: Island Press.

Costanza, R., L. Wainger, C. Folke, and K.-G. Mäler. 1993. Modeling complex ecological economic systems. *BioScience* 43 (8): 545-55.

Crissman, C.C., J.M. Antle, and S.M. Capalbo, eds. 1998. *Economic, environmental and health tradeoffs in agriculture: Pesticides and the sustainability of Andean potato production.* Boston: Kluwer Academic Publishers.

Cruise, J., and T.A. Lyson. 1990. Beyond the farmgate: Factors related to agricultural performance in two dairy communities. *Rural Sociology* 56 (1): 41-55.

Dairy Farmers of Ontario. 1998. *DFO Policies.* Mississauga: Dairy Farmers of Ontario.

Dalsgaard, J., C. Lightfoot, and V. Christensen. 1995. Towards quantification of ecological sustainability in farming systems analysis. *Ecological Engineering* 4: 181-89.

Davies, B., D. Eagle, and B. Finney. 1993. *Soil management.* 5th ed. Ipswich, UK: Farming Press Books and Videos.

Deitz, T., and P. Stern, eds. 2002. *New tools for environmental protection: Education, information and voluntary measures.* Washington, DC: National Research Council; National Academy Press.

De Lange, C.F.M. 1999. Nitrogen, phosphorus and potassium excretion with swine manure based on animal flow and feed usage. In *MCLONE4: An integrated systems approach to manure handling systems and nutrient management,* 39-42. Guelph: Manure Systems Research Group, University of Guelph.

De Lange, C.F.M., and V. Porteaux. 1999. Workshop on nutrient management. Banff Pork Seminar, Banff, AB.

Dent, J.B. 1993. Potential for systems simulation in farming systems research. In *Systems approaches for agricultural development,* ed. Penning de Vries et al., 325-39. Boston: Kluwer Academic Publishers.

Dent, J.B., G. Edward-Jones, and M.J. McGregor. 1995. Simulation of ecological, social and economic factors in agricultural systems. *Agricultural Systems* 49: 337-51.

Desjardins, R.L., and R. Riznek. 2000. Indicator: agricultural greenhouse gas budget. In *Environmental sustainability of Canadian agriculture: Report of the Agri-Environmental Indicator Project, a summary,* ed. T. McRae, C.A.S. Smith, and L.J. Gregorich, 14. Ottawa: Research Branch, Policy Branch, Prairie Farm Rehabilitation Administration, Agriculture and Agri-Food Canada.

Doyle, C.J. 1990. Application of systems theory to farm planning and control: Modelling resource allocation. In *Systems theory applied to agriculture and the food chain,* ed. J.G.W. Jones and P.R. Street, 89-112. London: Elsevier.

Driver, G., N. Moore, J. Schleihauf, G. Wall, J. Grevel, and R. Harkes. 1982. *Cropland soil erosion: Estimated cost to agriculture in Ontario.* Guelph: Ontario Institute of Pedology, University of Guelph.

Duff, S., D.P. Stonehouse, S.G. Hilts, and D.J. Blackburn. 1991. Soil conservation behavior and attitudes among Ontario farmers toward alternative government political responses. *Journal of Soil and Water Conservation* (May/June): 215-19.

Dumanski, J., and A.J. Smyth. 1994. The issues and challenges of sustainable land management. In *Proceedings of the International Workshop on Sustainable Land Management for the 21st Century. Volume 2: Plenary Papers,* ed. R.C. Wood and J. Dumanski, 11-21. Ottawa: Agriculture Institute of Canada.

Dumanski, J., H. Eswaran, and M. Latham. 1992. A proposal for an international framework for evaluating sustainable land management. In *Evaluation for sustainable land management in the developing world,* ed. J. Dumanski et al., 2: 25-45. Bangkok: IBSRAM.

Dumanski, J., V. Kirkwood, and P. Neave. 1995. Review and assessment of indicators for evaluating sustainable land management. *Canada Department of Agriculture Bulletin 95-07.* Ottawa: Canada Department of Agriculture and Agri-Food.

Edward-Jones, G., and M. McGregor. 1994. The necessity, theory, and reality of developing models of farm households. In *Rural and Farming Systems Analysis-European Perspectives,* ed. J.B. Dent and M.J. McGregor. Wallingford, UK: CAB International.

Ellis, F. 2000. *Rural livelihoods and diversity in developing countries.* Oxford: Oxford University Press.

Ellis, J.R., D.W. Hughes, and W.R. Butcher. 1991. Economic modeling of farm production and conservation decisions in response to alternative resource and environmental policies. *Northeastern Journal of Agricultural and Resource Economics* 20 (1): 98-108.

Ellstrand, N.C. 2003. *Dangerous liaisons? When cultivated plants mate with their wild relatives.* Baltimore: Johns Hopkins University Press.

Environmental Sciences Group. 2004. Trade-off analysis model, 29 January. Online at <http://www.tradeoffs.nl> (retrieved 7 April, 2004).

Eswaran, H., E. Pushparajah, and C. Ofori. 1993. Indicators and their utilization in a framework for evaluation of sustainable land management. Paper from Sustainable Agriculture and Rural Development Forum (SARD).

Evans, B., and K. Theobald. 2003. Local agenda 21 and the shift to "soft governance." In *Local environmental sustainability,* ed S. Buckingham and K. Thobald, 74-91. Cambridge, UK: Woodhead Publishing.

Executive Resource Group. 2001. Managing the environment: A review of best management practices. Executive summary. A report presented to the Government of Ontario.

Fairchild, G.L., D.A.J. Barry, M.J. Goss, A.S. Hamill, P. LaFrance, P.H. Milburn, R.R. Simard, and B.J. Zebarth. 2000. Groundwater quality. In *The health of our water: Toward sustainable agriculture in Canada,* ed. D.R. Coote and L.J. Gregorich, 61. Ottawa: Ministry of Public Works and Government Services Canada.

Felstehausen, H. 1993. Adjustment strategies of family farm operators: Case and system studies. In *Integrated Farming Systems Research Methods,* ed. E.A. Clark, 20-75. Guelph: University of Guelph.

Filson, G.C. 1983. Class and ethnic differences in Canadians' attitudes towards Native people's rights and immigration. *Canadian Review of Sociology and Anthropology* 20 (4): 454-82.

–. 1993. Comparative differences in Ontario farmers' environmental attitudes. *Journal of Agricultural and Environmental Ethics* 6 (2): 165-84.

–. 1996. Demographic and farm characteristic differences in Ontario farmers' views about sustainability policies. *Journal of Agricultural and Environmental Ethics* 9 (2): 165-80.

–. 1997. A farming systems approach to testing indicators of dairy farmers' quality of life and environmental attitudes and behaviour. FSR Research Project at the University of Guelph.

Filson, G.C., and M.A. McCoy. 1993. Farmers' quality of life: Sorting out the differences by class. *Rural Sociologist* 13 (1): 14-37.

Filson, G.C., and R.M. Friendship. 2000a. Waterloo Region and Perth County perceptions of hog production. In *Sustaining agriculture in the 21st century* (proceedings of the 4th biennial meeting, North American Chapter, International Farming Systems Association), ed. J.R. Ogilvie, J. Smithers, and E. Wall, 309-16. Guelph: University of Guelph.

–. 2000b. The public's perception of pork production. In *2000 Ontario Swine Research Review,* 3. Guelph: University of Guelph; OMAFRA; Ontario Pork.

Filson, G.C., C. Paine, and J.R. Taylor. 1999. Relating dairy farmers' quality of life to their environment. Paper presented at INDEX-99, St. Petersburg, Russia.

Filson, G.C., W.C. Pfeiffer, and J. Duncan-Robinson. 1998. A dairy portrait: Quality of life in our industry. *Ontario Milk Producer* 75 (7): 20-24.

Filson, G.C., W.C. Pfeiffer, C. Paine, and J.R. Taylor. 2003. The relationship between Grand River dairy farmers' quality of life and economic, social and environmental aspects of their farming system. *Journal of Sustainable Agriculture* 22 (1): 61-78.

Filson, G.C., D.P. Stonehouse, R. Rudra, S. Hilts, W. Caldwell, and C. Duke. 2001. Farming systems research. *Omafra Final Report.* Guelph: University of Guelph/OMAF Resources Management and the Environment Program, 29 January.

–. 2002. Policy incentives for soil and water conservation. *Final Report.* Guelph: University of Guelph/OMAF Resources Management and the Environment Program, 22 November.

Fitch, L., and B.W. Adams. 1998. Can cows and fish co-exist? *Canadian Journal of Plant Science* 78 (2): 191-98.

FitzGibbon, J., R. Plummer, and R. Summers. 2000a. Environmental Farm Plan indicators survey. Report for the Ontario Soil and Crop Improvement Association. Guelph: University of Guelph.

–. 2000b. Analysis of the impacts of implementation of the Environmental Farm Plan. Report for the Ontario Soil and Crop Improvement Association. Guelph: University of Guelph.

FitzGibbon, J., and L. Thacker. 2001. Nutrient management planning in Ontario: Preparedness of animal agriculture. Report of the Profiles in Animal Agriculture and Impacts of Regulation OMAFRA Special Projects Fund. Guelph: University of Guelph.

Fleming, R.A., B.A. Babcock, and E. Wang. 1998. Resource or waste? The economics of swine manure storage and management. *Review of Agricultural Economics* 20 (1): 96-113.

Flora, C. 1992. Building sustainable agriculture: A new application of farming systems research and extension. *Journal of Sustainable Agriculture* 2 (3): 37-50.

Forrester, J.W. 1968. *Principles of systems.* 2nd ed. Cambridge, MA: Wright Allen Press.

Frayne, Bruce. 2002. *Global food security, agriculture and environment.* Global and Geopolitical Food Issues: ENSC 315, Lecture 5. Online at <http://www.queensu.ca/envst> (retrieved 13 June 2003).

Friedmann, H. 1986. Family enterprises in agriculture: Structural limits and political possibilities. In *Agriculture: People and policies,* ed. G. Cox, P. Lowe, and M. Winter, 20-40. London: Allen and Unwin.

–. 2002. Locating rural sociology in the world economy, inter-state system, and biosphere. Paper presented at the Rural Studies of Canada: Critique and New Directions Conference, Guelph.

Friedmann, H., and P. McMichael. 1989. Agriculture and the state system: The rise and decline of national agriculture, 1870 to the present. *Sociologia Ruralis* 29 (2): 93-117.

Fulhage, C. 1994. Economics of two dairy manure management systems. In *Proceedings of the Liquid Manure Application Systems Conference,* ed. R. Koelsch, 65-75. Ithaca, NY: Northeast Regional Agricultural Engineering Service (NRAES-79).

Fuller, A.M. 1994. Sustainable rural communities in the arena society. In *Towards sustainable rural communities,* ed J. Bryden, L. Leblanc, and C. Teal, 133-39. Guelph: The Guelph Seminar Series, University of Guelph School of Rural Planning and Development.
–. 2002. A Canadian rural development perspective. Talk given at the University of Guelph, 28 March.
Gallopin, G. 1995. The potential of agroecosystem health as a guiding concept for agricultural research. *Ecosystem Health* 1 (3): 129-40.
Galt, D. and T. Barrett. Summary of consultations. *Task Force on Intensive Agricultural Operations in Rural Ontario Consultation.* Online at <http://www.gov.on.ca/OMAFRA/english/infores/releases/081603_b.html> (retrieved 6 June 2002).
Gertler, M.E. 1999. Sustainable communities and sustainable agriculture on the prairies. In *Communities, development, and sustainability across Canada,* ed. J.T. Pierce and A. Dale. Vancouver: UBC Press.
Gibbon, D. 1994. Farming systems research/extension. In *Rural and farming systems and analysis – European Perspectives,* ed. J.B. Dent and M.J. McGregor. Wallingford, UK: CAB International.
Giles, J.L., and M. Dalecki. 1988. Rural well-being and agricultural change in two farming regions. *Rural Sociology* 53 (1): 40-55.
Gillespie, B. 2001. Growth of the organic sector and its potential contribution to the development of sustainable agriculture: An extension perspective. MSc major paper, University of Guelph.
Goddard, E., A. Weersink, C. Kevin, and C.G. Turvey. 1993. Economics of structural change in agriculture. *Canadian Journal of Agricultural Economics* 41: 475-89.
Goldschmidt, W. 1978a. *As you sow: Three studies in the social consequences of agribusiness.* Montclair: Allanheld, Osmun.
–. 1978b. Large-scale farming and the rural social structure. *Rural Sociology* 43 (Fall): 362-66.
Gomes, E. 2000. Article review of pork articles appearing in three newspapers from January 1995 to December 1999. Unpublished manuscript, University of Guelph.
Gordon, R., M. LeClere, P. Schuepp, and R. Brunke. 1988. Field estimates of ammonia volatilization from swine manure by a simple micrometeorological technique. *Canadian Journal of Soil Science* 68: 369-80.
Goss, K.F. 1979. Review of Goldschmidt, As you sow. *Rural Sociology* 44: 802-6.
Goss, M.J. 1994. Biophysical criteria for the evaluation of intensive cropping and livestock management systems. In *Proceedings of the International Workshop on Sustainable Land Management for the 21st Century. Volume 2: Plenary Papers,* ed. R.C. Wood and J. Dumanski, 189-201. Ottawa: Agriculture Institute of Canada.
Goss, M.J., and D.A. Barry. 1995. Ground water quality: Responsible agriculture and public perceptions. *Journal of Agricultural and Environmental Ethics* 8: 52-64.
Goss, M.J., D.A.J. Barry, and D.L. Rudolph. 2000. Ontario farm groundwater quality survey. In *The health of our water: Toward sustainable agriculture in Canada,* ed. D.R. Coote and L.J. Gregorich, 64. Ottawa: Ministry of Public Works and Government Services Canada.
Goss, M.J., D.P. Stonehouse, and J.G. Giraldez, eds. 1996. *Managing for dairy and swine: Towards developing a decision support system.* Guelph: Centre for Land and Water Stewardship, University of Guelph.
Gunderson, L.H., and C.S. Holling, eds. 2002. *Panarchy: Understanding transformation in human and natural systems.* Washington, DC: Island Press.
Hadley, G. 1964. *Nonlinear and dynamic programming.* Reading, MA: Addison-Wesley.
Hall, A. 1997. Sustainable agriculture and neoliberalism. Paper presented at the annual meeting of the Rural Sociological Society, Toronto.
Häni, F. 2001. Natural Regulation at the farm level: Key factor for a holistic approach to sustainability. International Scientific Seminar of the Swiss National Science Foundation and Ministry of Science and Technology of the People's Republic of China: Plant Production with Sustainable Agriculture, 28 May-1 June, Zhuhai, China.
Häni, F., E. Boller, and S. Keller. 1998. Natural regulation at the farm level. In *Enhancing biological control,* ed. C.H. Pickett and R.L. Bugg. Berkeley: University of California Press.

Hardaker, J.B., R.B.M. Huirne, and J.R. Anderson. 1997. *Coping with risk in agriculture*. Wallingford, UK: CAB International.
Harker, D.B., P.A. Chambers, A.S. Crowe, G.L. Fairchild, and E. Kienholz. 2000. Understanding water quality. In *The health of our water: Toward sustainable agriculture in Canada,* ed. D.R. Coote and L.J. Gregorich, 27-42. Ottawa: Ministry of Public Works and Government Services Canada.
Harrington, M. 1992. Measuring sustainability. *Journal of Farming Systems Research-Extension* 3 (1): 1-20.
Harwood, R.R. 1990. A history of sustainable agriculture. In *Sustainable Agricultural Systems,* ed. C. Edwards, R. Lal, P. Madden, R. Miller, and H.R. House, 3-19. Ankeny, IA: Soil and Conservation Society.
Hazell, P.B.R., and R.D. Norton. 1986. *Mathematical programming for economic analysis in agriculture*. New York: Macmillan.
Healey, S. 2003. Rural social movements and the prospects for sustainable rural communities: Evidence from Bolivia. PhD qualifying paper, University of Guelph.
Health Canada. Overview of review of Recombinant Bovine Somatotropin (rbST) by Health Canada. Online at <http://www.hc-sc.gc.ca/english/media/releases/1998/9875bke2.htm> (retrieved 7 April 2004).
Hildebrand, P.E. 1986. Economic characteristics of small-scale, limited resource family farms. In *Perspectives on farming systems research and extension,* ed. P.E. Hildebrand, 57-59. Boulder, CO: Lynne Rienner.
Huffman, E. 2000. Indicator: Soil cover by crops and residue. In *Environmental sustainability of Canadian agriculture: Report of the Agri-Environmental Indicator Project, a summary,* ed. T. McRae, C.A.S. Smith, and L.J. Gregorich, 4. Ottawa: Research Branch, Policy Branch, Prairie Farm Rehabilitation Administration, Agriculture and Agri-Food Canada.
Ignizio, J.P., and T.M. Cavalier. 1994. *Linear programming*. Englewood Cliffs, NJ: Prentice Hall.
Ikerd, J.E. 1993. The need for a systems approach to sustainable agriculture. *Agriculture, Ecosystems and Environment* 46: 147-60.
Izacs, A.-M.N., and M.J. Swift. 1994. On agricultural sustainability and its measurement in small-scale farming in sub-Saharan Africa. *Ecological Economics* 11: 105-25.
Jeffrey, S.R., and M.D. Faminow. 1995. Programming models: Potential applications to agricultural marketing research. In *Prices, products and people: Analyzing agricultural markets in developing countries,* ed. G.J Scott, 439-60. Boulder, CO: Lynne Rienner.
Jeffrey, S.R., R.R. Gibson, and M.D. Faminow. 1992. Nearly optimal linear programming as a guide to agricultural planning. *Agricultural Economics* 8: 1-19.
Johnson, P. 1996. Keynote address. Great Lakes Agriculture Summit, Kellog Center for Continuing Education, Michigan State University, 23-24 April.
Johnson, S., and J. Claar. 1986. FSR/E: Shifting the intersection between research and extension. *Agricultural Administration* 21: 81-93.
Jordan, A., and O'Riordan, T. 1999. The precautionary principle: Contemporary environmental policy and politics. In *Public health and the environment: Implementing the precautionary principle,* ed. C. Raffensperger and J. Tickner, 13-35. Washington, DC: Island Press.
Joseph, A.E., J.M. Lidgard, and R. Bedford. 2001. Dealing with ambiguity: On the interdependence of change in agriculture and rural communities. *New Zealand Geographer* 57 (1): 16.
Joy, D., S. Bonte-Gelok, and C. Merkley. 2000. Surface water quality in rural Ontario. In *The health of our water: Toward sustainable agriculture in Canada,* ed. D.R. Coote and L.J. Gregorich, 53. Ottawa: Ministry of Public Works and Government Services Canada.
Kay, J., and E. Schneider. 1992. Thermodynamics and measures of ecological integrity. In *Ecological indicators: Problems and approaches,* ed. S.A. Lewin, M.A. Harwell, J.R. Kelly, and K.D. Kimball. New York: Springer-Verlag.
Keddie, P., and J. Wandel. 2001. *Animal agriculture in Ontario 1996. A profile of its sectors and patterns of distribution: An atlas resource*. Guelph: University of Guelph.
Kelland, D.L., and D.P. Stonehouse. 1984. *Mixed integer programming model for analyzing manure-handling system alternatives*. Guelph: School of Agricultural Economics and Extension Education, University of Guelph.

Kelly, K. 1996. Keynote address. Great Lakes Agriculture Summit, Kellog Center for Continuing Education, Michigan State University, April 23-24.
King, D.J., G.C. Watson, G.J. Wall, and B.A. Grant. 1994. The effects of livestock manure application and management on surface water quality. Summary Technical Report – GLWQP, Agriculture and Agri-Food Canada. London, ON: Pest Management Research Centre, Agriculture and Agri-Food Canada.
Kingwell, R.S., and D.J. Pannell, eds. 1987. *MIDAS, a bioeconomic model of a dryland farm system.* Wageningen, Netherlands: Pudoc.
Kirchoff, S., B.G. Colby, and J. LaFrance. 1997. Evaluating the performance of benefit transfer: An empirical inquiry. *Journal of Environmental Economics and Management* 33 (2): 75-93.
Klupfel, E.J. 2000. OMAFRA special projects proposal on greenfuels. Submitted by the University of Guelph.
–. 2001. Moving towards a crop-based economy. PhD qualifying examination paper, University of Guelph.
Klupfel, E.J., and G.C. Filson. 2000. Family roles on Grand River dairy farms. In *Sustaining agriculture in the 21st century* (proceedings of the 4th biennial meeting, North American Chapter, International Farming Systems Association), ed. J.R. Ogilvie, J. Smithers, and E. Wall, 301-8. Guelph: University of Guelph.
Kneen, B. 1989. *From land to mouth: Understanding the food system.* Toronto: NC Press.
–. 1990. *Trading up: How Cargill, the world's largest grain company, is changing Canadian agriculture.* Toronto: NC Press.
–. 1999. *Farmageddon: Food and the culture of biotechnology.* Gabriola Island, BC: New Society.
Koroluk, R., D. Culver, A. Lefebrve, and T. McRae. 1995. Management of farm nutrient and pesticide inputs. In *Environmental sustainability of Canadian agriculture: Report of the Agri-Environmental Indicator Project, a summary,* ed. T. McRae, C.A.S. Smith, and L.J. Gregorich, 41-54. Ottawa: Research Branch, Policy Branch, Prairie Farm Rehabilitation Administration, Agriculture and Agri-Food Canada.
–. 2000. Indicator: Management of farm nutrient and pesticide inputs. In *Environmental sustainability of Canadian agriculture: Report of the Agri-Environmental Indicator Project, a summary,* ed. T. McRae, C.A.S. Smith, and L.J. Gregorich, 5. Ottawa: Research Branch, Policy Branch, Prairie Farm Rehabilitation Administration, Agriculture and Agri-Food Canada.
Krug, K.L. 2000. Linking urban and rural contexts in Niagara to build a more sustainable agriculture system. In *Sustaining agriculture in the 21st century* (proceedings of the 4th biennial meeting, North American Chapter, International Farming Systems Association), ed. J.R. Ogilvie, J. Smithers, and E. Wall, 275-83. Guelph: University of Guelph.
Kulshreshtha, S.N., M. Boehm, M. Bonneau, and J.C. Giraldez. 1999. *Canadian economic and emissions model for agriculture: Report 1 model description.* Ottawa: Policy Branch, Agriculture and Agri-Food Canada.
Lambert, D.K., and B.A. McCarl. 1985. Risk modeling using direct solution of nonlinear approximations of the utility function. *American Journal of Agricultural Economics* 67: 846-52.
Law, A.M., and W.D. Kelton. 1982. *Simulation modeling and analysis.* New York: McGraw-Hill.
Laxer, G. 1995. Social solidarity, democracy and global capitalism. *Canadian Review of Sociology and Anthropology* 32 (3): 287-314.
Le Heron, R. 1993. *Globalized agriculture: Political choice.* New York: Pergamon Press.
Le Heron, R., and B. van der Knapp. 1995. Industrial spaces as contexts for human resource development. In *Human resources and industrial spaces: A perspective on globalization and localization,* ed. B. van der Knapp and R. Le Heron, 3-27. Toronto: John Wiley and Sons.
Lightfoot, C., and R. Noble. 1993. A participatory experiment in sustainable agriculture. *Journal of Farming Systems Research-Extension* 4 (1): 11-34.
Lightfoot, C., P.T. Dalsgaard, M.P. Bimbao, and F. Fermin. 1993. Farmer participatory procedures for managing and monitoring sustainable farming systems. *Journal of the Asian Farming Systems Association* 2 (2): 67-87.
Lintner, A., and A. Weersink. 1999. Endogenous transport coefficients: Implications for improving water quality from multi-contaminants in an agricultural watershed. *Environmental and Resource Economics* 14 (2): 269-96.

Lobao, L. 2002. Political economy of agriculture and rural America during the 1990s: An overview. Paper presented at the Rural Studies of Canada: Critique and New Directions Conference, Guelph.

Lyson, T.A., and G.W. Gillespie. 1995. Producing more milk on fewer farms: Neoclassical and neostructural explanations of changes in dairy farming. *Rural Sociology* 60 (3): 493-504.

Ma, K.K.Y., and J.R. Ogilvie. 1998. MCLONE3: A decision support system for management of liquid dairy and swine manure. In *Computers in agriculture* (proceedings of the 7th international conference, 26-30 October), ed. F.S. Zazueta and X. Jiannong, 480-86. St. Joseph, MI: ASAE.

MacDonald, J. 1999. *Dairy statistical handbook, 1997-98.* 14th ed. Mississauga: Dairy Farmers of Ontario.

MacDonald, K.B. 2000a. Indicator: Risk of water contamination by nitrogen. In *Environmental sustainability of Canadian agriculture: Report of the Agri-Environmental Indicator Project, a summary,* ed. T. McRae, C.A.S. Smith, and L.J. Gregorich, 12. Ottawa: Research Branch, Policy Branch, Prairie Farm Rehabilitation Administration, Agriculture and Agri-Food Canada.

–. 2000b. Indicator: Residual nitrogen. In *Environmental sustainability of Canadian agriculture: Report of the Agri-Environmental Indicator Project, a summary,* ed. T. McRae, C.A.S. Smith, and L.J. Gregorich, 16. Ottawa: Research Branch, Policy Branch, Prairie Farm Rehabilitation Administration, Agriculture and Agri-Food Canada.

MacGregor, R.J., and T. McRae. 2000. Driving forces affecting the environmental sustainability of agriculture. In *Environmental sustainability of Canadian agriculture: Report of the Agri-Environmental Indicator Project, a summary,* ed. T. McRae, C.A.S. Smith, and L.J. Gregorich, 3-4. Ottawa: Research Branch, Policy Branch, Prairie Farm Rehabilitation Administration, Agriculture and Agri-Food Canada.

MacGregor, R.J., R. Lindenbach, S. Weseen, and A. Lefebvre. 2000. Indicator: Energy use. In *Environmental sustainability of Canadian agriculture: Report of the Agri-Environmental Indicator Project, a summary,* ed. T. McRae, C.A.S. Smith, and L.J. Gregorich, 171-77. Ottawa: Research Branch, Policy Branch, Prairie Farm Rehabilitation Administration, Agriculture and Agri-Food Canada.

Manure Systems Research Group. 1999. *MCLONE4: An integrated systems approach to manure handling systems and nutrient management.* Guelph: University of Guelph.

Marten, G.G. 1988. Productivity, stability, sustainability, equitability and autonomy as properties for agroecosystem assessment. *Agricultural Systems* 26: 291-316.

Martin, L.A. 1997. An introduction to feedback. Paper No. D-4691. Prepared for the Massachusetts Institute of Technology System Dynamics in Education Project. Online at <http://sysdyn.clexchange.org/road-maps/rm-toc.html> (retrieved 21 October 2003).

Martins, O., and T. Dewes. 1992. Loss of nitrogenous compounds during composting of animal wastes. *Bioresource Technology* 42: 103-11.

Marx, K. 1959. *Capital: A critique of political economy,* vol. 3. Moscow: Progress Publishers.

McBride, R.A., P.J. Joosse, and G. Wall. 2000. Indicator: Risk of soil compaction. In *Environmental sustainability of Canadian agriculture: Report of the Agri-Environmental Indicator Project, a summary,* ed. T. McRae, C.A.S. Smith, and L.J. Gregorich, 10. Ottawa: Research Branch, Policy Branch, Prairie Farm Rehabilitation Administration, Agriculture and Agri-Food Canada.

McCoy, M.A., and Filson, G.C. 1996. Working off the farm: Impacts on quality of life. *Social Indicators Research* 37 (2): 149-63.

McCullum, J., D. Rapport, and M. Miller. 1995. *Assessing agroecosytem health via the soil subsystem: A functional approach.* Guelph: Faculty of Environmental Sciences, University of Guelph.

McRae, T., and C.A.S. Smith. Regional analysis of environmentally sustainable agriculture. In *The health of our water: Toward sustainable agriculture in Canada,* ed. D.R. Coote and L.J. Gregorich, 181-96. Ottawa: Ministry of Public Works and Government Services Canada.

McRae, T., C.A.S. Smith, and L. Gregorich, eds. 2000. *Environmental sustainability of Canadian agriculture: Report of the Agri-Environmental Indicator Project.* Ottawa: Agriculture and Agri-Food Canada.

McSweeny, W.T., and J.S. Shortle. 1989. Reducing nutrient application rates for water quality protection on intensive livestock areas: Policy implications of alternative producer behavior. *Northwestern Journal of Agricultural and Resource Economics* 18 (1): 1-11.

Minguez, M.I., C. Romero, and J. Domingo. 1988. Determining fertilizer combination through goal programming with penalty functions: An application to sugar beet production in Spain. *Journal of the Operational Research Society* 39: 61-70.

Moore, K. 1995. The conceptual basis for targeting farming systems: Domain, zones, and typologies. *Journal of Farming Systems Research-Extension* 5 (2): 19-38.

Neave, P., E. Neave, T. Weins, and T. Riche. Availability of wildlife habitat on farmland. In *Environmental sustainability of Canadian agriculture: Report of the Agri-Environmental Indicator Project, a summary,* ed. T. McRae, C.A.S. Smith, and L.J. Gregorich, 145-56. Ottawa: Research Branch, Policy Branch, Prairie Farm Rehabilitation Administration, Agriculture and Agri-Food Canada.

Norman, D.W. 2002. *The farming systems approach: A historical perspective.* Paper presented at the 17th symposium of the International Farming Systems Association, Lake Buena Vista, FL.

O'Connor, D. 2002. *Report of the Walkerton Inquiry: A strategy for safe drinking water.* Toronto: Walkerton Inquiry Commission.

O'Connor, J. 1994. Is sustainable capitalism possible? In *Is capitalism sustainable? Political economy and the politics of ecology,* ed. M. O'Connor. New York: Guilford Press.

ODFAP (Ontario Dairy Farm Accounting Project). 1999. *Annual Report 1998.* Guelph: Canadian Dairy Commission; Dairy Farmers of Ontario; Ontario Ministry of Agriculture, Food and Rural Affairs; University of Guelph.

Ogilvie, J.R., D.A. Barry, M.J. Goss, and D.P. Stonehouse. 2000. Balancing environmental and economic concerns in manure management by use of an on-farm computerized decision support system, MCLONE4. In *Proceedings of the 8th international symposium on animal, agriculture and food processing wastes.* Des Moines, IA: American Society of Agricultural Engineers.

O'Halloran, I.P. 1993. Ammonia volatilization from liquid hog manure: Influence of aeration and trapping systems. *Soil Science Society of America Journal* 57: 1300-3.

Okey, B. 1995. *Building a conceptual basis for agroecosystem health: Systems approaches and properties.* Guelph: Agroecosystem Health Project, Faculty of Environmental Sciences, University of Guelph.

OFEC (Ontario Farm Environmental Coalition). 1993. *Ontario Environmental Farm Plan.* Toronto: Ontario Farm Environmental Coalition.

OMAF (Ontario Ministry of Agriculture and Food). 2003a. *10 steps to complete a nutrient plan for livestock and poultry manure.* Toronto: Queen's Printer.

–. 2003b. *New products from OMAF.* Toronto: Queen's Printer. Online at <http://www.gov.on.ca/OMAFRA/english/products/newpubs.html> (retrieved 25 June 2003).

–. 2003c. *Nutrient Management Act and Regulations.* Toronto: Queen's Printer. Online at <http://www.gov.on.ca/OMAFRA/english/agops/> (retrieved 23 June 2003).

–. 2003d. *Ontario environmental farm plan.* Toronto: Queen's Printer. Online at <http://www.gov.on.ca/OMAFRA/english/environment/efp/efp.htm> (retrieved 14 June 2003).

OMAFRA (Ontario Ministry of Agriculture, Food and Rural Affairs). 1994a. *Field crop production: Best management practices.* Guelph: Ontario Ministry of Agriculture, Food and Rural Affairs.

–. 1994b. *Soil Management: Best management practices.* Guelph: Ontario Ministry of Agriculture, Food and Rural Affairs.

Owen, L., W. Howard, and M. Waldron. 2000. Conflicts over farming practices in Canada: The role of iterative conflict resolution approaches. *Journal of Rural Studies* 16: 475-83.

Pannell, D.J. 1997. Sensitivity analysis of normative economic models: Theoretical framework and practical strategies. *Agricultural Economics* 16 (2): 139-52.

Pannell, D.J., and S. Schilizzi. 1999. Sustainable agriculture: A question of ecology, equity, economic efficiency or expedience? *Journal of Sustainable Agriculture* 13 (4): 57-66.

Pannell, D.J., L.R. Malcolm, and R.S. Kingwell. 2000. Are we risking too much? Perspectives on risk in farm modeling. *Agricultural Economics* 23 (1): 69-78.

Paris, Q. 1991. *An economic interpretation of linear programming.* Ames: Iowa State University Press.
Patten, L.H., J.B. Hardaker, and D.J. Pannell. 1988. Utility-efficient programming for whole-farm planning. *Australian Journal of Agricultural Economics* 32: 88-97.
Paul, J.W., and E.G. Beauchamp. 1993. Nitrogen availability for corn in soils amended with urea, cattle slurry, and solid and composted manures. *Canadian Journal of Soil Science* 73: 253-66.
Paul, J.W., E.G. Beauchamp, and X. Zhang. 1993. Nitrous and nitric oxide emissions during nitrification and dentrification from manure-amended soil in the laboratory. *Canadian Journal of Soil Science* 73: 539-53.
Peet, Mary. 2001. *Conservation tillage.* Raleigh, NC: North Carolina State University. Online at <http://www.cals.ncsu.edu/sustainable/peet/tillage/tillage.html> (retrieved 23 June 2003).
Petheram, R.J. 1986. Farming/systems research at BPT: Some progress and constraints. In *Proceedings of the workshop on farming systems research and development, Central Research Institute for Animal Sciences, Ciawi-Bogor, Indonesia, March 1985,* ed. J. Levine and M. Sabrani, 32-46. Research Institute for Animal Protection.
Pezzey, J. 1992. *Sustainable development concepts: An economic analysis.* Environment Paper 2. Washington, DC: World Bank.
Pfeiffer, W.C., and G.C. Filson. 1999. What future challenges await the dairy industry? Managerial attitudes point the way. In *Sustaining agriculture in the 21st century* (proceedings of the 4th biennial meeting, North American Chapter, International Farming Systems Association), ed. J.R. Ogilvie, J. Smithers, and E. Wall, 111-20. Guelph: University of Guelph.
Philips, W.E., and T.S. Veeman. 1987. Alternative incentives and institutions for water and soil conservation. *Canadian Water Resource Journal* 12 (3): 27-33.
Phillips, D. 2001. A changing climate: What's up with the weather? Paper presented at a workshop on risks and opportunities for the agricultural sector from climate change, University of Guelph.
Pierzynski, G.M., J. Thomas Sims, and George F. Vance. 1994. *Soils and environmental quality.* Boca Raton, FL: Lewis Publishers.
Pingali, P.L., and P. Rogers, eds. 1995. *Impact of pesticides on farmer health and the rice environment.* Boston: Kluwer Academic Publishers.
Potapchuk, W.R., J.P. Crocker, D. Boogaard, W.H. Schechter. 1998. *Building community: Exploring the role of social capital and local government.* Washington, DC: Program for Community Problem Solving.
Qui, Z., and T. Prato. 1998. Economic evaluation of riparian buffers in an agricultural watershed. *Journal of the American Water Resources Association* 34 (4): 877-90.
Rae, A.N. 1994. *Agricultural management economics: Activity analysis and decision making.* Wallingford, UK: CAB International.
Raffensperger, C., and J. Tickner, eds. 1999. *Protecting public health and the environment: Implementing the precautionary principle.* Washington, DC: Island Press.
Rapport, D. 1994. The concept of agroecosystem health and its applications to agriculture. In *Agroecosystem Health: Proceedings of an International Workshop,* ed. O. Nielson. Guelph: University of Guelph.
–. 1995. Ecosystem services and management options as blanket indicators of ecosystem health. *Journal of Aquatic Ecosystem Health* 4: 97-105.
Rapport, D., and A. Friend. 1979. *Towards a comprehensive framework for environmental statistics: A stress-response approach.* Cat. No. 11-510. Ottawa: Statistics Canada.
Redclift, M. 1987. *Sustainable development: Exploring the contradictions.* London: Routledge.
Rehman, T., and C. Romero. 1984. Multiple-criteria decision making techniques and their role in livestock ration formulation. *Agricultural Systems* 15: 23-49.
Richmond, L.A., G.C. Filson, C. Paine, W.C. Pfeiffer, and J.R. Taylor. 2000. Non-farm rural Ontario residents' perceived quality of life. *Social Indicators Research* 50 (2): 159-186.
Ritchie, J.T., and J.B. Dent. 1994. Data requirements for agricultural systems: Research and applications. In *Systems approaches for agricultural development,* ed. Penning de Vries et al. Dordrecht, Netherlands: Kluwer Academic Publishers.

Ritter, W.F. 1989. Odor control of livestock wastes: State-of-the-art in North America. *Journal of Agricultural Engineering Research* 42: 51-62.

Roka, F.M., and D.L. Hoag. 1996. Manure value and liveweight swine decisions. *Journal of Agricultural and Applied Economics* 28 (1): 193-202.

Röling, N.G. 1985. Extension science: Increasingly preoccupied with knowledge systems. *Sociologia Ruralis* 25 (3/4): 269-89.

Romero, C., and T. Rehman. 1989. *Multiple criteria analysis for agricultural decisions*. Amsterdam: Elsevier Science Publishers.

Rotz, C.A. and U.S. Gupta. 1995. *DAFOSYM for Windows*. US Dairy Forage Research Centre, Research Summaries. Online at <http://www.dfrc.wisc.edu/RS95_pdfs/fp6.pdf/> (retrieved 15 June 2003).

Rotz, C.A., D.R. Buckmaster, D.R. Mertens, and J.R. Black. 1989. DAFOSYM: A dairy forage system model for evaluating alternatives in forage conservation. *J. Dairy Science* 72: 3050-63.

Rudolph, D.L., D.A.J. Barry, and M.J. Goss. 1998. Contamination in Ontario farmstead domestic wells and its association with agriculture. 2. Results from multilevel monitoring well installations. *Journal of Contaminant Hydrology* 32: 295-311.

Ryan, T. 1999. The Clean Up Rural Beaches Program: Environmentalism in action? MSc thesis, University of Guelph, Rural Extension Studies.

Saha, B. 2003. Changes in farm structure and rural communities: The dynamics of interactions and implications for sustainability of rural communities. University of Guelph Rural Studies PhD Program Qualifying Paper.

Samson, R., A. Weil, A. Archinstall, and J. Quin. 1992. Manure management in conservation farming system: A research final report. Quebec: Resource Efficient Agricultural Production.

Schaller, N. 1990. Mainstreaming low-input agriculture. *Journal of Soil and Water Conservation* 45: 9-23.

Schwarzweller, H. 1996. *Dairy farm restructuring in America and Australia: Dilemma of the middle*. Ann Arbor: Department of Sociology, Michigan State University.

Scoones, I. 1998. *Sustainable rural livelihoods: A framework for analysis*. Brighton, UK: Institute for Development Studies, University of Sussex.

Sellen, D., W. Howard, and E. Goddard. 1993. Production to consumption systems research: A review of methods and approaches. Report prepared for the International Development Research Centre, Ottawa, by the Department of Agricultural Economics and Business, University of Guelph.

Serman, N., and G.C. Filson, 2000. Factors affecting farmers' adoption of soil and water conservation. In *Sustaining agriculture in the 21st century* (proceedings of the 4th biennial meeting, North American Chapter, International Farming Systems Association), ed. J.R. Ogilvie, J. Smithers, and E. Wall, 69-78. Guelph: University of Guelph.

Sharpley, A.N., and A.D. Halvorson. 1994. The management of soil phosphorus availability and its impact on surface water quality. In *Advances in soil science: Soil processes and water quality,* ed. R. Lal and B.A. Stewart, 7-90. Boca Raton, FL: CRC Press.

Shelton, I.J., et al. 2000. Indicator: Risk of water erosion. In *Environmental sustainability of Canadian agriculture: Report of the Agri-Environmental Indicator Project, a summary,* ed. T. McRae, C.A.S. Smith, and L.J. Gregorich, 6. Ottawa: Research Branch, Policy Branch, Prairie Farm Rehabilitation Administration, Agriculture and Agri-Food Canada.

Simon, H.A. 1957. *Models of man: Social and rational*. New York: Wiley.

Skipton, S., and D. Hay. 1998. *Drinking water nitrate and methemoglobinemia ("blue baby" syndrome)*. Lincoln, NE: Cooperative Extension, Institute of Agriculture and Natural Resources, University of Nebraska. File G1369 under Water Resource Management. Online at <http://www.ianr.unl.edu/pubs/water/g1369.htm> (retrieved 30 June 2003).

Smit, B. 2001. How is climate change relevant to farmers? Paper presented at a workshop on risks and opportunities from climate change for the agricultural sector, for the agricultural sector from climate change, University of Guelph.

Smit, B., and J. Smithers. 1993. Sustainable agriculture: Interpretations, analyses and prospects. *Canadian Journal of Regional Science* 16 (3): 499-524.

Smit, B., D. Waltner-Toews, D. Rapport, E. Wall, G. Wichert, and E. Gwyn. 1997. *Agroecosystem health: Analysis and assessment.* Guelph: University of Guelph.

Smith, V.K., and J.C. Huang. 1995. Can markets value air quality? A meta-analysis of hedonic property value models. *Journal of Political Economy* 103 (2): 209-27.

Smith, V.K., G. Van Houtven, and S. Pattanayak. 1999. Benefit transfer as preference calibration. Discussion paper for Resources for the Future, Washington, DC.

Smith, W., and L. Saunders. 1995. Agricultural policy reforms and sustainable land management: A New Zealand case study. *Australian Geographer* 26: 112-18.

Smithers, J. 1997. Towards an integrated framework for assessing the sustainability of farming systems in the province of Ontario. Working paper for the University of Guelph Farming Systems Research Project.

Smithers, J., and A. Joseph. 2000. Agricultural and rural community change in Ontario: Understanding complementarity and conflict. In *Sustaining agriculture in the 21st century* (proceedings of the 4th biennial meeting, North American Chapter, International Farming Systems Association), ed. J.R. Ogilvie, J. Smithers, and E. Wall, 265-74. Guelph: University of Guelph.

Smithers, J., E. Wall, and C. Swanton. 2002. An integrated framework for solving problems in sustainable agriculture. *Journal for Farming Systems Research-Extension* 7 (2): 43-58.

Soil Conservation Society of America. 1983. *Soil erosion: Its agricultural and environmental implications for southwestern Ontario.* Don Mills, ON: Soil Conservation Society of America.

Sommer, S.G., H. Mikkelsen, and J. Mellgvist. 1995. Evaluation of meteorological techniques for measurements of ammonia loss from pig slurry. *Agricultural and Forest Meteorology* 74: 169-79.

Spoelstra, S.F. 1980. Origin of objectionable odorous components in piggery wastes and the possibility of applying indicator components for studying odour development. *Agriculture and the Environment* 5: 241-60.

Statistics Canada. 1997. *Historical overview of Canadian agriculture.* Ottawa: Ministry of Industry.

Steingraber, S. 1997. *Living downstream: An ecologist looks at cancer and the environment.* Reading, MA: Addison-Wesley.

Stockle, C., R. Papendick, K. Saxton, G. Campbell, and F. Van Evert. 1994. A framework for evaluating the sustainability of agricultural production systems. *American Journal of Alternative Agriculture* 9: 45-50.

Stonehouse, D.P., and M.J. Goss. 1999. A decision support system for combined technical, economic, and environmental components of manure handling. In *Manure management 99* (proceedings of a tri-provincial conference on manure management, 22-25 June), 107. Saskatoon: Saskatchewan Agriculture and Food.

Surgeoner, G. 1996. Keynote presentation, Great Lakes Agriculture Summit, Kellog Center for Continuing Education, Michigan State University, 23-24 April.

Swanson, L.E. 1990. Rethinking assumptions about farm and community change. In *American rural communities,* ed. A.E. Luloff and L.E. Swanson. Boulder, CO: Westview Press.

Swiss Agency for the Environment. Forests and landscape, biodiversity: Promoting partnerships with the private sector to save our global heritage. Online at <http://www.umwelt-schweiz.ch/imperia/md/content/buwalcontent/folder/03-06-03tagderumwelt/13.pdf> (retrieved 7 April 2004).

Tisdall, P. 1992. Approaches to sustainable agriculture: Seven case studies. Discussion paper from the Science Council of Canada, Ministry of Supply and Services, Ottawa.

Tomasi, T., K. Segerson, and J. Braden. 1994. Issues in the design of incentive schemes for nonpoint source pollution control. In *Nonpoint source pollution regulation: Issues and analysis,* ed. C. Dosi and T. Tomasi. Boston: Kluwer Academic Publishers.

Toombs, M. 1997. The rising concern in rural Ontario regarding swine production. In *Swine production and the environment: Living with our neighbours,* OMAFRA Staff, 1-4. Guelph: OMAFRA.

Troughton, M. 1998. The countryside in Ontario: An editorial. In *The countryside in Ontario: Evolution, current challenges and future directions,* ed. M. Troughton and J.G. Nelson, 3-5. Waterloo, ON: Heritage Resources Centre, University of Waterloo.

–. 2002. Enterprises and commodity chains. In *The sustainability of rural systems: Geographical interpretations,* ed. I.R. Bowler, C.R. Bryant, and C. Cocklin. Boston: Kluwer Academic Publishers.

Troughton, M.J. 1982. Progresss and response in the industrialization of agriculture. In *The Effects of Modern Agriculture on Rural Development,* ed. G. Enyedi and I. Volgyes. New York: Pergamon Press.

Tuitock, J.K., L.G. Young, B.J. Kerr, and C.F.M. de Lange. 1993. Digestible amino acid pattern for growing finishing pigs fed practical diets. *Journal of Animal Science* 71 (Suppl. 1): 167.

US National Library of Medicine. 1995. Hazardous substances databank. Online at <http://extoxnet.orst.edu/pips/ghindex.html> (retrieved 13 June 2003).

Van der Ploeg, J.D. 1995. From structural development to structural involution: The impact of new development in Dutch agriculture. In *Beyond modernization: The impact of endogenous rural development,* ed. J.D. van der Ploeg and G. van Dijk, 219-32. Assen: Van Gorcum.

van Evert, Frits. Modeling whole-farm production and environment impacts. *Wageningen University and Research Centre, the Netherlands.* Online at <http://www.plant.wageningen-ur.nl/projects/modeling-framework/whole-farm/whole-farm-2003.html> (retrieved 20 July 2003).

Veeraraghavan, S. 1985. The role of farm organizations. In *Farming and the rural community in Ontario,* ed. A.M. Fuller. Toronto: Foundation for Rural Living.

Wall, E.S. 2002. *Sustainable rural communities in an era of globalization: Reporting on research in rural Ontario.* Guelph: University of Guelph.

Wall, E.S., A. Weersink, and C. Swanton. 1998. Ontario and ISO 14000. Report prepared for the Ontario Farm Environmental Committee and the Ontario Federation of Agriculture. Guelph: University of Guelph.

Waltner-Toews, D. 1996. Ecosystem health: A framework for implementing sustainability in agriculture. *BioScience* 46 (9): 686-89.

Waltner-Toews, D., and E. Wall. 1997. Emerging perplexity: In search of post-normal questions for agroecosystem health. *Social Science and Medicine* 45 (11): 1741-49.

Weber, G. 1996. Heterogeneity and complexity in farming systems: Towards an evolutionary perspective. *Journal for Farming Systems Research-Extension* 6 (2): 15-32.

Weersink, A., C. Nicholson, and J. Weerhewa. 1998a. Multiple job holding among dairy farm families in New York and Ontario. *Agricultural Economics* 18: 127-43.

Weersink, A., J. Livernois, J. Shogren, and J. Shortle. 1998b. Economic instruments and environmental policy in agriculture. *Canadian Public Policy* 24 (3): 309-27.

Weersink, A., R. McKitrick, and M. Nailor. 2001. Voluntary cost share programs: Lessons from economic theory and their application to rural water quality programs. *Current Agricultural, Food and Resource Issues* 2: 23-36.

Winson, A. 1992. *The intimate commodity: Food and the development of the agro-industrial complex in Canada.* Toronto: Garamond Press.

Woodhill, J., and N. Röling. 1998. The second wing of the eagle: The human dimension in learning our way to more sustainable futures. In *Facilitating sustainable agriculture,* ed. N. Röling and M.A.E. Wagemakers, 46-71. Cambridge: Cambridge University Press.

Woodley, S., J. Kay, and G. Francis, eds. 1993. *Ecological integrity and the management of ecosystems.* Delray Beach, FL: St. Lucie Press.

World Commission on Environment and Development (WCED). 1987. *Our common future.* Oxford: Oxford University Press.

Wotowiec, P., and P. Hildebrand. 1988. Research, recommendation and diffusion domains: A farming systems approach to targeting. In *Gender issues in farming systems research and extension,* ed. S.V. Poats, M. Schmink, and A. Spring, 73-86. Boulder, CO: Westview Press.

Wu, J., and K. Segerson. 1995. The impact of policies and land characteristics on potential groundwater pollution in Wisconsin. *American Journal of Agricultural Economics* 77 (5): 1033-47.

Yiridoe, E., and A. Weersink. 1997. A review and evaluation of agroecosystem health analysis: The role of economics. *Agricultural Systems* 55 (4): 601-26.

–. 1998. Marginal abatement costs of reducing groundwater N pollution from intensive and extensive management choices. *Agricultural and Resource Economics Review* 27 (2): 169-85.

Yiridoe, E., R.P. Voroney, and A. Weersink. 1997. Impact of alternative farm management practices on nitrogen pollution of groundwater: Validation and application of CENTURY model. *Journal of Environmental Quality* 26 (5): 1255-63.

Zander, P., and H. Kachele. 1999. Modeling multiple objectives of land use for sustainable development. *Agricultural Systems* 59 (3): 311-25.

Newspaper and Magazine Articles

Avery, R. 2003. Walkerton group irate over delay in manure controls. *Toronto Star,* 23 March, A4.

Bourette, S. 2000. Deadly *E. coli* traced to cattle. *Globe and Mail,* 4 August.

Chase, S. 2002. Canada's dairy subsidies deemed illegal. *Globe and Mail,* 25 June, B1.

Chase, S., M. MacKinnon, and P. Brethour. 2003. Martin goes cool on Kyoto. *Globe and Mail,* 3 December, A1, A5.

Cobb, C. 2001. Canadians wary of genetically modified foods: Also against cloning: But minor risks are acceptable in disease control, study says. *National Post,* 2 January, A6.

Contenta, S. 2003. "Canada can learn" from UK errors. *Toronto Star,* 24 May, A25.

Egan, K. 1997. Neighbours raise a stink over piggery: Couple upset about firm's plan to house 800 pregnant sows next door. *Ottawa Citizen,* 13 June, D7.

Flack, R. 1998. Commercial feed industry seeks tough regulations. *Toronto Star,* 19 June, A23.

Gallon, G. 2000. The real Walkerton villain. *Globe and Mail,* 20 December.

Ho, Q., D.P. Stonehouse, and M.J. Goss. 2000. Manure: farmers ready to let computers help. *Ontario Farmer,* 15 August.

Khimji, M. 1998. Intensive pig life is a tortuous one. *Toronto Star,* 29 May, A25.

Kilpatrick, K. 2000. Concern grows about pollution from megafarms: Fight against huge factory operations gains attention as manure suspect in outbreak. *Globe and Mail,* 30 May, A9.

Laidlaw, S. 2003. Complacency is real killer in mad cow. *Toronto Star,* 25 May, A1.

Landau, S. 2000. EC beef study tarnishes Canadian farmers' image. *Globe and Mail,* 21 November.

Lawton, V. 2000. Ottawa aims to cut greenhouse gases. *Toronto Star,* 6 October, A6.

Lindgren, A. 1999. Resort areas raise stink about factory hog farms: Cottagers, neighbours fear beaches could be polluted. *Ottawa Citizen,* 23 September, C3.

MacCharles, T. 2001. Ontario farm labour law struck down. *Toronto Star,* 21 December, A2.

Mittelstaedt, M. 2000. Waterworks found deficient. *Globe and Mail,* 8 December.

Nikiforik, A. 2000. When water kills. *Maclean's,* 12 June, 18.

Priddle, R. 2001. Climate change goes to market. *International Herald Tribune,* 3 August, 6.

Shecter, B. 1998. Big sky is in hog heaven. *Financial Post,* 11 June, 32.

Toronto Star. 2003. Mad cow disease. 21 May, A7.

Walkom, T. 2003. This was a problem just waiting to happen. *Toronto Star,* 22 May, A6.

Whalen, S. 2003. Milk farmers on the ropes after ruling cuts off exports. *Toronto Star,* 27 May, A8.

Contributors

Dean A. Barry is a postdoctoral fellow, Land Resource Science, University of Guelph.

G.W. de Vos is Commodity Market and Market Revenue Insurance Economist, Policy and Farm Finance Division, Ontario Ministry of Agriculture and Food, Guelph.

Chris Duke is Environmental Management Specialist, Applied Research, Agricultural and Rural Division, Ontario Ministry of Agriculture and Food.

Glen C. Filson is Associate Professor, School of Environmental Design and Rural Development, and former FSR Director, University of Guelph.

John FitzGibbon is Director, School of Environmental Design and Rural Development, University of Guelph, and Chair of the Ontario Environmental Farm Coalition.

Michael J. Goss is Professor and Chair of Land Stewardship and Research, University of Guelph, and Director for the Resources Management and Environment Program, University of Guelph, Ontario Ministry of Agriculture and Food Partnership for Research.

Scott Jeffrey is Associate Professor, Department of Rural Economy, University of Alberta.

Georgina Knitel is Vice President Marketing, Cox Financial Group, Lethbridge, Alberta.

Murray H. Miller is University Professor Emeritus, Land Resource Science, University of Guelph.

John R. Ogilvie is University Professor Emeritus, School of Engineering, University of Guelph.

Santiago Olmos is a research fellow, Institute of Rural Sciences, University of Wales, Aberystwyth, United Kingdom.

David Pannell is Associate Professor, Department of Agricultural and Resource Economics, University of Western Australia.

Wayne C. Pfeiffer is Associate Professor, Agricultural Economics and Business, University of Guelph.

Ryan Plummer is Assistant Professor, Department of Recreation and Leisure Studies, Brock University.

John Smithers is Associate Professor of Geography, University of Guelph.

D.P. Stonehouse is Professor, Agricultural Economics and Business, University of Guelph.

Clarence Swanton is Professor and Chair, Plant Agriculture, University of Guelph.

Robert Summers is a doctoral candidate, School of Environmental Design and Rural Development, University of Guelph.

Ellen Wall is a sociologist with Canadian Climate Impacts and Adaptation Research Network

Alfons Weersink is Professor, Agricultural Economics and Business, University of Guelph.

Index

Printed and bound in Canada by Friesens
Set in Stone by Artegraphica Design Co. Ltd.
Copy editor: Frank Chow
Proofreader: Sarah Munro
Indexer: Annette Lorek